D0860277

ENZYMES

THE M.I.T. PAPERBACK SERIES

ENZYMES

J. B. S. HALDANE

THE M.I.T. PRESS
MASSACHUSETTS INSTITUTE OF TECHNOLOGY
CAMBRIDGE, MASSACHUSETTS

First M.I.T. Press Paperback Edition, August, 1965

Library of Congress Catalog Card Number: 65-23929
Printed in the United States of America

PREFACE TO THE PAPERBACK EDITION

A REPRINT such as this serves five purposes. In the first place, it enables those who are interested in a topic to find out what was known about it at some date in the past. An enzymologist should be aware of what wa and what was not known in 1930. Secondly, it may describe lines of worl which have not been followed up systematically since it was written an might be fruitfully pursued once more. Tables II, V, and XII sugge generalisations which may turn out to be false or to throw a good deal light on enzyme action in general. Had I retired from active research, whi I have not, and were I living in a country with good library facilities, should have tried to expand these tables into a book. Thirdly, it may gi references to researches which have been wholly forgotten but which m now prove to be relevant to modern problems. Fourthly, it can serve as Awful Warning. How, readers may ask, could Haldane, whose accoun enzyme kinetics in Chapters III and V requires many additions, but corrections, back so many wrong horses as he did in some of the l r chapters? Perhaps some modern molecular biologists may seem equ y shortsighted thirty-four years hence. And lastly, such a reprint may conv e the author that he has not wasted his life.

I think the main advances in enzymology since I wrote this book e been three. First, in the purification of numerous enzymes, starting th Sumner's pioneer work. All turn out to be proteins, and many lack a s-thetic group. Secondly, in a few cases we know, and in the rest we ess with high probability, that the structure of an enzyme is exactly sp ied by that of one or more genes. Finally, we know that metabolic pro es, such as oxidations and syntheses, are usually series of enzyme-cat sed reactions, each reversible because it causes a small free energy chang and involving small molecules, often nucleotides such as the adenosine hos-phates and the phosphopyridine nucleotides, which were unknown i 930. In that year I confidently hoped that thirty years hence we shoul now the exact structure of the active centres of a number of enzymes l be able to calculate their properties from this structure. This hope has yet been fulfilled, partly because the active centres seems to include am acid residues in several adjacent and loosely bonded peptide chains. It is one of my ambitions to live long enough to read a paper whose author explains, from the amino-acid sequences of two enzymes, why one must be an esterase and the other a peptidase.

Finally, I want to remedy an omission. As Sir Frederick Hopkins was not only a very modest man, but editor of the series in which this book first appeared, I could not state how much I owed him. Until I came to work in his laboratory I had only used enzymes as analytical tools and had not thought about them for five consecutive minutes. Hopkins convinced me that they were a central topic in biochemistry. My father, J. S. Haldane, had shown about 1910 that though haemoglobin is a large molecule, its reactions can be predicted from the laws known to hold for small molecules. I had only to bring the ideas of these two great men together to produce an account of enzyme action which, though sketchy, seems to have been largely correct.

J. B. S. HALDANE

Genetics and Biometry Laboratory, Bhubaneswar,
Orissa, India, May, 1964.

PREFACE

THIS book differs in two respects from the other members of the series to which it belongs. It lays no claim to completeness. For an exhaustive account of the subject the student is referred to Oppenheimer and Kuhn's monumental *Die Fermente*, while a very full treatment of many aspects may be found in Euler's *Chemie der Enzyme*. I do not propose to treat of the biology of enzymes, either from the point of view of their function or formation; still less of their applications in the laboratory or factory. Even so the account of their purely chemical side will be inadequate.

Secondly, the subject has already been treated in this series by the late Sir Wm. Bayliss. In view of the great additions to our knowledge in the last few years a demand has arisen for a new work on the subject. *The Nature of Enzyme Action*, like all Bayliss' work, was strongly individual, and could not have been brought up to date without at least a partial loss of that individuality. Moreover, Bayliss' book was to some extent a polemic for the view now universally accepted, that enzymes are catalysts, and for the thermodynamical implications of such a view, which are not always so fully recognized. As such it is in no sense superseded by the present book, which, to a large extent, attempts to build on the ground cleared by Bayliss. In order to keep the book within reasonable dimensions, I have been forced to assume a considerable knowledge of organic and physical chemistry in its readers, but I have added an appendix to Chapter VI on some recent work on carbohydrate chemistry which is not yet wholly incorporated into the textbooks. I have dealt in a very summary manner with the enzymes concerned in the complicated processes of alcoholic fermentation, and respiration. It will only be possible to study them in a really quantitative way when they have been separated from the other catalysts with which they cooperate. Moreover, the subject of alcoholic fermentation has already been dealt with in this series by Prof. Harden, while Dr. Dixon is preparing a volume on the catalysis of oxidations.

This book is based on a course of lectures which I have delivered here since 1923. It owes much to conversations with Drs. Dixon, Quastel, and Wurmser, and with Mr. Woolf.

J. B. S. HALDANE

Department of Biochemistry,
Cambridge University, April, 1930.

CONTENTS.

CHAPTER I.

INTRODUCTION.

Enzymes as Catalysts.

An enzyme may be defined as a soluble, colloidal, organic catalyst, produced by a living organism. This rules out, on the one hand, crystalloidal catalysts such as glutathione; and on the other, the active surfaces which appear to be responsible for many oxidations and reductions, but of which the active portions have not yet been got into solution. Although these latter will be considered in this volume, they will not be dealt with fully. Many enzymes can still act when thrown out of solution either by precipitation or adsorption, yet all the substances usually designated as such can be brought into solution. The definition will be out of date when an enzyme is prepared synthetically, but this book will be out of date long before that time.

A few enzymes are to be found in natural secretions, such as digestive juices and milk. The large majority can only be obtained from the interior of cells, and since Buchner's [1897] preparation of zymase from yeast it has become clear that they are responsible for many intracellular reactions. Among their properties are intense activity, specificity, and intimate dependence on hydrogen ion concentration, temperature, and other conditions.

An idea of the activity may be gained from the fact that Willstätter, Schneider and Wentzel's [1926] most active preparation [1] of invertase hydrolysed six times its weight of cane-sugar per second at 15° C., whilst Euler and Josephson's [1927, 1] best catalase [1] broke up two hundred and twenty times its weight of H_2O_2 per second at 0° C., and Willstätter and Pollinger's [1923] best peroxidase [1] activated one thousand times its weight of H_2O_2 per second (to oxidize leuco-malachite green) at 20° C. Specificity may be absolute or quantitative. Thus zymase, which ferments *d*-glucose, is quite inactive with *l*-glucose; on the other hand, pancreatic lipase hydrolyses ethyl *d*-mandelate, although the hydrolysis of its optical antipode is more rapid. [2]

[1] See p. 172. [2] See Chap. VI.

Similarly rabbits can oxidize and store *l*-mannose, though more slowly than *d*-mannose, according to Neuberg and Mayer [1903]; although yeast cannot ferment the *l*-sugars. The dependence on external conditions will be described later. Suffice it to say that a careful regulation of pH and temperature is essential in all quantitative work on enzyme chemistry.

Catalysis.

A catalyst can only promote a reaction provided that a loss of free energy results. This is somewhat less sweeping than the statement that a catalyst can merely accelerate a reaction which would otherwise occur, though the latter statement may also be true. But this does not mean that a catalyst merely alters the time-scale of events which would otherwise take place, or enables them to occur at a lower temperature. Side reactions are the rule in organic chemistry. Consider a substance A, of which, in the absence of a catalyst, 99 per cent. is converted into B, and 1 per cent. into C. If a catalyst accelerates the latter process a millionfold, 99·99 per cent. will be converted into C, and only ·01 per cent. into B. Hence the fate of A is qualitatively altered. Enzymes do not merely accelerate metabolic processes, they determine their direction. Thus, according to the enzymes present, glucose can be broken down to lactic acid on the one hand, or alcohol and CO_2 on the other.

But where any given reaction is concerned, an enzyme, like any other catalyst, cannot alter the final state of equilibrium, and if it accelerates any reaction, it must accelerate the reverse action to about the same extent. These two propositions have been proved for emulsin by Bayliss [1913], and by Bourquelot [1914] and his colleagues for emulsin and maltase, in a number of cases where the equilibrium can easily be reached from both sides; and Barendrecht [1920], Mack and Villars [1923], and Kay [1923] have shown that even in the case of such an apparently irreversible reaction as the hydrolysis of urea, the reverse action is catalysed by urease.

Hence this book is concerned with chemical kinetics, not with statics. The facts that, for example, many glucosides and bioses have been synthesised by means of enzymes by Bourquelot and others, and that around pH 4 Borsook and Wasteneys [1925] have synthesized proteins from peptone by means of pepsin are important. But they merely exemplify, on the one hand, the fact that enzymes are catalysts; on the other, the chemical statics of the particular systems concerned.

The key to a knowledge of enzymes is the study of reaction velocities, not of equilibria.

Some Alleged Exceptions.

Various cases occur in the literature in which an enzyme appears to affect the equilibrium reached, or not to accelerate the two opposing reactions to an equal extent. Thus if urease is added to an unbuffered urea solution, hydrolysis stops at a certain point. On adding more enzyme the reaction proceeds further, and so on. But in a buffered solution the reaction proceeds very nearly to completion in either case. In the absence of buffers the solution rapidly becomes alkaline, and the reaction ceases. The enzyme solution contains proteins and other buffers, and when more is added the alkalinity is diminished and the reaction can again proceed. Such " false equilibria " may be of importance biologically. Thus Dakin and Dudley [1913] showed that the formation of lactic acid from methyl-glyoxal by glyoxalase was inhibited by the lactic acid so formed ; and continued when alkali is added. This may explain the formation of lactic acid when living tissues are rendered alkaline, as by Anrep and Cannan [1923].

A more famous stumbling-block is furnished by the work of Dietz [1907] who found that when a certain mixture of isoamyl butyrate and water is saponified by picric acid, equilibrium is reached at 14·5 per cent. hydrolysis ; while pancreatic lipase hydrolyses 25 per cent. The equilibria were true, as they could be reached from both sides, and the rates of approach to the enzymatic equilibrium were equally affected by changes in the enzyme concentration, which did not affect the equilibrium itself. Bayliss [1925] ascribes this to adsorption of the reactants by the enzyme preparation. If so, however, the equilibrium should have varied with the amount of the enzyme preparation. It seems more likely that whereas in the strongly acid mixture the butyric acid was mainly present as such, in the neutral enzyme solution much of it was ionized, and that butyrate ions do not take part in the reaction :

$$\text{Butyric acid} + \text{isoamyl alcohol} \rightleftharpoons \text{isoamyl butyrate} + \text{water}.$$

If so, it is natural that hydrolysis would proceed further in such a case. The experiments should be repeated with due regard to the effects of pH, which may, of course, affect the equilibrium, as in the case of peptic hydrolysis of proteins. This explanation is supported by the work of Schlesinger [1926, 1927] who found that when strong acids were used as catalysts, either acids or neutral salts favoured synthesis of ethyl acetate. By extrapolation he was able to calculate the equilibrium in presence of small amounts of acid, but even in ·05M acid concentration the equilibrium constant might be reduced by 25 per cent. below this value. It is probable, therefore, that the picric acid rather than the lipase was the cause of the deviations found by Dietz. The case of the hydrolysis of starch, where equilibrium may be dependent on the amount of enzyme, will be considered in Chapter VII.

Again Armstrong [1905] held that maltase catalyses the series of reactions : maltose + $H_2O \rightarrow 2$ glucose $\rightarrow$ isomaltose + H_2O, but not the reverse series, while emulsin catalyses the reverse reactions only. He has since [1924] abandoned this view. Another apparent exception is the case of hexose-diphosphoric acid formation by yeast zymase. This contains an enzyme which, with the co-operation of cozymase, forms (2,5)-fructose di-phosphate from hexoses and inorganic phosphates, and another enzyme, phosphatase, which hydrolyses it to hexose and phosphate. But here the first reaction is only one half of a coupled reaction :—

$$2C_6H_{12}O_6 + 2KH_2PO_4 \rightarrow 2C_3H_6O_3 + C_6H_{10}O_4(KHPO_4)_2 + 2H_2O,$$

and presumably some of the free energy liberated in the formation of the triose molecules goes to build up the unstable hexose-diphosphate. (Cf. Harden, 1923.)

If enzymes do not act as mere catalysts, a possibility exists of obtaining supplies of free energy in defiance of the second law of thermodynamics. As we have seen, all enzymatic reactions so far studied obey this law. Nevertheless, if anything analogous to a Maxwell demon exists outside textbooks it presumably has about the dimensions of an enzyme molecule, and hence researches which show that the second law holds in the case of enzyme action possess a very general interest.

Some Examples of Non-enzymatic Catalysis.

Inorganic and crystalloidal catalysts on the whole exhibit less specificity and less dependence on pH than do enzymes, and the latter fact has led many writers to suppose that the peculiar sensitivity of enzymes to pH is due to their colloidal nature. This is not necessarily the case, as follows from the experiments of Widmark and Jeppson [1922] and Ljungren [1925] on the decarboxylation of acetoacetic acid by aniline and a variety of other bases. Acetoacetic acid decomposes into acetone and carbon dioxide about fifty times as fast as the acetoacetate ion. On varying the pH the rate varies with the degree of dissociation of the acetoacetic acid. If now aniline be added it is found that the maximum rate of decarboxylation occurs at a relatively low hydrogen ion concentration. The extra CO_2 production, which varies with the aniline concentration, is always a maximum at pH 4·2. Now, since the dissociation constants of aniline and acetoacetic acid are $10^{-4.8}$ and $10^{-3.7}$, the quantity of aniline acetoacetate formed is a maximum at just this pH. Everything is therefore explained on the assumption that aniline acetoacetate breaks down much more quickly than the free acid, and a variety of data agree quantitatively with this point of view. A number of other bases act in a similar manner, each with its characteristic pH optimum. Thus, ammonia, glycine, and propylamine have optima near pH 7. The chemical theory of enzyme action postulates that an enzyme behaves in just this manner. For example the chemical compound of invertase and sucrose is held to unite with a molecule of water many millions of times more rapidly than does free sucrose. Euler and Josephson [1927] have found the closest parallelism between the effect of substrate concentration on the destruction of H_2O_2 by hæmin and by liver catalase, and conclude that the type of catalyst-substrate union is the same in both cases.

Fajans [1910] gave an example of homogeneous catalysis exhibiting quantitative specificity between two optical isomers analogous to that of the lipases; *d*- and *l*-camphocarboxylic acids yield camphor and carbon dioxide in accordance with the equation

$$C_8H_{14}\left<\begin{array}{l}CH.COOH\\ |\\ CO\end{array}\right. = C_8H_4\left<\begin{array}{l}CH_2\\ |\\ CO\end{array}\right. + CO_2$$

when heated in solution in a variety of solvents. In optically inactive media such as aniline and acetophenone the two break up at equal rates, but in nicotine the *d*-acid is more rapidly decarboxylated. Thus at 70° the rates were in the ratio of 1·12 : 1. Similarly a solution of quinidine *d*-camphocarboxylate in acetophenone broke up 1·46 times as fast as quinidine *l*-camphocarboxylate. This is natural enough, as the two substances are not mirror images and doubtless differ in their physical properties, solubilities, and so forth. It is a reasonable deduction that enzymes which have different actions on optically antipodal substances are themselves composed of asymmetric molecules.

Of course inorganic colloidal catalysts themselves vary in their efficiency with hydrogen ion concentration. Thus Bredig and von Berneck [1899] found that addition of NaOH to a platinsol first raised and then lowered its activity in decomposing hydrogen peroxide. I do not, however, know of any such experiments in which the pH was accurately determined.

The Measurement of Reaction Velocity.

In all problems of enzyme chemistry (as opposed to applications of enzymes to synthesis or other external goals) we are concerned with the measurement of a reaction velocity. It is unfortunately too often assumed that the velocity so measured is that of the catalysed reaction. In many cases diffusion may be the limiting factor. Thus Henri [1906] found that when saccharase was immobilized in gelatin and placed in the solution of its substrate, the temperature coefficients of the reaction catalysed by it was reduced from Q_{10} 1·9 to 1·2. This is natural enough, since the substrate had to diffuse to and from the enzyme over relatively large distances, and the temperature coefficient of a diffusion is nearly unity. The same consideration probably accounts for the very low temperature coefficients often found in the case of lipases, which commonly act in heterogeneous systems. In many cases the velocity cannot be measured adequately. In the case of proteases and amylases a large number of processes are occurring successively, and different results are obtained according to which we measure. Thus Northrop [1922, 4] obtained quite different results as to the effect of substrate concentration on the tryptic digestion of casein according as he measured the rate of disappearance of casein or that of amino nitrogen liberation. The velocity generally falls off as a reaction proceeds, so it is always advisable to compare the times taken to effect a given change, rather than the changes occurring in a given time. This is of course essential in the case of a qualitative change, such as the coagulation of milk, the loss of its iodine reaction by starch, or the complete decolorization of methylene blue. It is fortunate that in many cases the initial velocity of reaction remains constant for a relatively long period, and can therefore be accurately measured. A measurement of the initial velocity avoids complications introduced by destruction of the enzyme by its substrate (e.g. H_2O_2), or by unfavourable temperature or pH, or its reversible inactivation by the products of the reaction, as well as by the effects of diminishing substrate concentration and of the back reaction. As long as the velocity remains constant it may usually be assumed that none of these processes is occurring to a measurable extent.

The velocity is generally, but not always, found to be proportional to the enzyme concentration. As the substrate concentration is increased, the velocity generally at first increases proportionally to it, and later reaches an asymptotic value. From such data deductions have been drawn as to the nature of the enzyme-substrate union. Heating has two distinct effects: a reversible increase in the reaction velocity, and an irreversible destruction of the enzyme. Hydrogen ions may influence the reaction reversibly either by affecting the rate of change of the enzyme-substrate compound (using the word to include adsorbates), or the rate at which it is formed. In all cases there appears to be an optimum pH, though this may cover a wide range. Hydrogen ions may also influence a reaction irreversibly by destroying the enzyme. Inorganic or organic substances may favour the reaction, generally by combining with the enzyme. They may retard it, or "poison" the enzyme, in several different ways. Perhaps the most important distinction lies between those which do and do not compete with the substrate. The percentage depression of activity by the latter class is independent of the amount of substrate. That due to the former class diminishes as the amount of substrate is increased. A study of such inhibitors has thrown much light both on the chemical nature of enzymes, and of their union with their substrate.

Great advances have been made of recent years in the purification of enzymes. Although only three enzymes at most have been obtained in an approximately pure state, yet a great deal of light has been thrown by these researches on the chemical and physical nature of enzymes. Moreover, several enzymes, such as trypsin, exhibit greater specificity when purified than had previously been attributed to them, whilst the failure to fractionate others affords evidence of their unitary character. When an enzyme is believed to be pure its action can be tested on a variety of substrates, and still further light is thus thrown on the nature of the enzyme-substrate union. Nevertheless it is clear that enzyme chemistry has only just begun, and that most of its conclusions are only provisional.

History.

At the end of the eighteenth century various kinds of fermentation had been distinguished, but nothing serious was known concerning the agents causing them, save for Spallanzani's [1783] discovery that meat was liquefied by the gastric juice of hawks. The history of enzymes perhaps begins with Planche's [1810, 1820] discovery that extracts of plant roots will blue guaiacum tincture, and he called the thermolabile soluble agent cyanogen, although Gay-Lussac had already described C_2N_2. The next step was Robiquet and Boutron-Chalard's [1830] discovery of the hydrolysis of amygdalin, which they had isolated, by bitter almonds. Liebig and Wöhler [1837] and Robiquet [1838] investigated the properties of the active principle concerned and named it emulsin. Meanwhile Leuchs [1831] described the diastatic action of ptyalin, Payen and Persoz [1833] showed that malt diastase could be precipitated by alcohol, Fauré [1835] described sinigrinase and Schwann [1836] pepsin. Probably Planche should be regarded as having discovered the first enzyme. In 1837 Berzelius included fermentation under catalytic processes, and Liebig in a series of papers (e.g. 1870) developed a purely chemical theory of enzyme action. Meanwhile, however, it had been recognized that yeast was alive, and Pasteur (e.g. 1871) showed that living yeast was necessary for fermentation. He had already shown [Pasteur, 1858] that in the fermentation of ammonium *dl*-tartrate, only the *d*-acid is destroyed. This laid the foundation of our knowledge of specificity.

Pasteur's work led to a distinction between "unorganized ferments," such as pepsin and diastase, and "organized ferments," such as yeast and the lactic acid producing bacteria. These had not been distinguished by former workers. For this reason the word "enzyme" was for some time preferable to "ferment" as a name for the latter. As, however, no one to-day would describe a bacillus as a ferment, there seems to be little objection to the use of this word for enzymes. The Pasteur-Liebig controversy was almost terminated by Buchner's [1897] discovery that a cell-free yeast extract can cause alcoholic fermentation; though this is still occasionally denied (e.g. Kostychev, Medvedev, and Kardo-Sysoyeva, 1927, but cf. Kluyver and Struyk, 1927). Pasteur had been right in all the facts which he adduced, nevertheless Liebig was correct in believing that fermentation was possible in the absence of life. On the other hand, certain catalyses, e.g. glycolysis by erythrocytes, appear to be dependent on cell structure. In 1862 Danilewsky separated pancreatic trypsin and amylase by adsorption of the former on collodion. Both enzymes were obtained almost protein-free, and amylase was shown by dialysis to be a colloid. But these remarkable results were not followed up.

The idea was widely held that the various enzymes which had meanwhile been prepared owed their activity to a residue of vital force from the living cell. It was with this question in view that O'Sullivan and Tompson [1890] made the first investigation of the action of an enzyme (yeast saccharase) by really quantitative methods. They concluded that the course of inversion was that of a unimolecular action. It was not till 1898 that Duclaux showed that this was not the case. In 1894 Emil Fischer and his colleagues began the series of investigations on which our ideas of specificity are based. These were the logical outcome of his determination of the molecular structure of sugars and peptides. The former work was extended by his pupil E. F. Armstrong, who showed [Armstrong, 1904] that specificity extends to inhibition by compounds related to the substrate. Another pupil, Abderhalden, continued the work on polypeptides. In 1898 Croft Hill performed the first enzymatic synthesis, that of maltose. Meanwhile Battelli and Stern [summary in 1912], and Bach with Chodat and others [summary, 1910], began the systematic study of oxidases, and distinguished many of the more important enzymes of this class. They worked on foundations laid largely by Schönbein, Traube, and Bertrand [see Battelli and Stern, 1912].

All work on enzymes was, however, at the mercy of unanalysable changes in environmental conditions until Sörensen [1909] pointed out the dependence of enzyme activity on pH. This was rapidly followed up by Michaelis and others. (See Chap. II.) At the same time the ideas of Brown [1902] and Henri [1903] with regard to the union of enzyme and substrate were developed by Michaelis and Menten. (See Chap. III.) The study of many of the oxidizing-reducing enzymes was enormously simplified by Thunberg's [1917] invention of the methylene blue technique, which has since yielded most important results in his hands and those of others.

Since that time the subject has enormously developed. The literature will be found in the later chapters. Among the more important centres of research activity have been Munich, Stockholm, Cambridge, Berlin, Paris, and New York. In Munich, Willstätter and his colleagues, especially Waldschmidt-Leitz and Kuhn, have greatly contributed to our knowledge of the mechanism of enzyme action, and to the methods of separating and purifying enzymes. A somewhat similar field was covered by Euler and his colleagues, Josephson, Myrbäck and others in Stockholm. In Cambridge, Quastel, Dixon and their colleagues studied the catalysis of oxidations and reductions. In New York, Northrop studied the kinetics of the proteases, while Nelson and his colleagues made measurements of the utmost precision of the kinetics of saccharase action. In Berlin,

Neuberg and his colleagues studied in detail the behaviour of a variety of different individual enzymes derived from yeast and other sources, Pringsheim the enzymatic hydrolysis of polysaccharides in relation to their molecular structure, Meyerhof and his colleagues the enzymes of muscle, and Rona the action of lipases. In Paris Bertrand, Bourquelot, Bridel and others studied the glucosidases, especially from the point of view of synthesis.

Important contributions to our knowledge of enzymes are now being made in most civilized countries, and no adequate summary is possible. The most important lines of attack are (1) Purification of enzymes. (2) Study of their specificity both as regards substrates and inhibitors. (3) Study of the kinetics of the catalysed reactions under different circumstances. (4) Investigation of the catalysed processes from the point of view of structural and physical chemistry. (5) Physiology of enzymes, including a study of cell-models containing enzymes along with other substances. (6) Study of the conditions under which enzymes are formed, e.g. influence of nutrients, and biochemical genetics. The last two topics are hardly touched upon in this book. A full history of enzyme chemistry would, of course, include a survey of its relations with pure chemistry. Thus enzyme chemistry owes very much to Hofmeister's theory of protein structure and Fischer's of sugar structure. On the other hand, but for the use of enzymes as reagents, tryptophan could hardly have been discovered, maltose would be a rare substance, and the complex glucosides could hardly be classified. It is probable that the use of enzymes as specific reagents in organic chemistry is still in its infancy.

CHAPTER II.

THE INFLUENCE OF ENZYME CONCENTRATION AND HYDROGEN ION CONCENTRATION.

Enzyme Concentration.

As a general rule the velocity of the reaction varies with the enzyme concentration. Thus Hudson [1908], on varying the concentration of yeast saccharase eightfold, found the following percentages of cane sugar inverted at 30° C., and optimal pH :—

TABLE I.

Saccharase Concn.	Time.	Initial Sucrose Per Cent.		
		4·55.	9·09.	27·3.
2·00	15	73·2	45·3	11·2
1·50	20	73·2	44·8	11·2
1·00	30	72·9	45 3	11·5
0·50	60	72·9	45·2	11·4
0·25	120	73·1	45·2	10·9

In each the product of the saccharase concentration and the time was equal, as was the amount of work done by the enzyme. Very numerous results with other enzymes generally agree with the same law. The majority of the results in the literature which disagree with it are due to the effect of the enzyme preparation on the pH of the solution. This is of course avoided by working with buffers. The most notable real exceptions are to be found among the proteases. Many authors have found that the velocity of digestion was roughly proportional to the square root of the concentration of pepsin. Northrop [1920, 2] showed that this was more or less true for impure pepsin, but with a highly purified product it was no longer the case (Fig. 1). On addition of products of peptic digestion to the pepsin the velocity fell off in high enzyme concentrations owing to the presence of inhibitors which combined reversibly with the pepsin. The question is dealt with in Chapter III. Similar deviations from linearity found in the case of amylase will be discussed in Chapter VII.

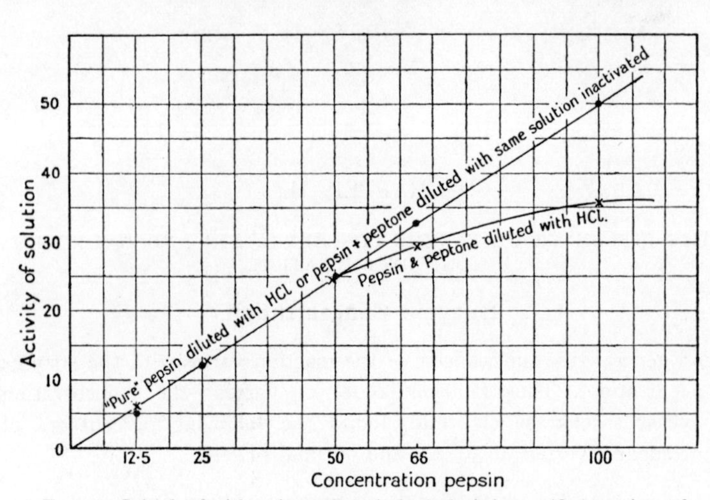

FIG. 1.—Initial velocities of egg albumin hydrolysis by purified pepsin, and pepsin + peptone. Crude preparations behave like the latter.

[Northrop, 1920, 2.]

Influence of pH on Reaction Velocity.

An unfavourable pH may cause destruction of an enzyme. If, then, we are to estimate the reversible effects on reaction velocity

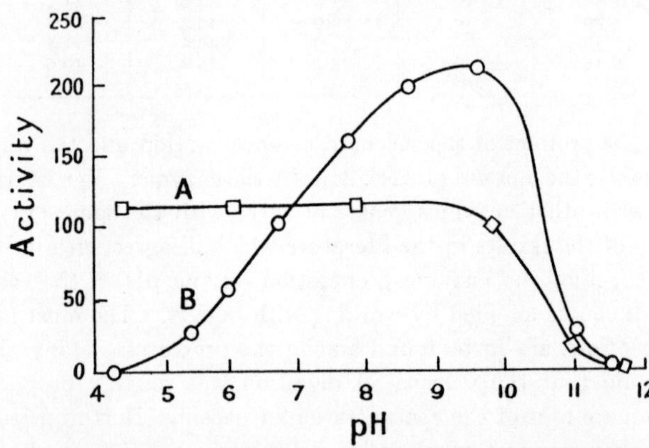

FIG. 2.—The influence of hydrogen ion concentration on the activity of the tyramine oxidase system.

A = Activity of enzyme at pH 7·3 after being subjected for 5 mins. to various hydrogen ion concentrations.

B = Activity of enzyme at different hydrogen ion concentrations.

The apparent optimum is clearly due to irreversible inactivation.

[Hare, 1928.]

this must be avoided or allowed for. Yeast saccharase, for example, undergoes fairly rapid irreversible inactivation at pH below 3. In order to allow for this it is desirable to measure the initial velocities of reaction due to a strong enzyme solution when working in this

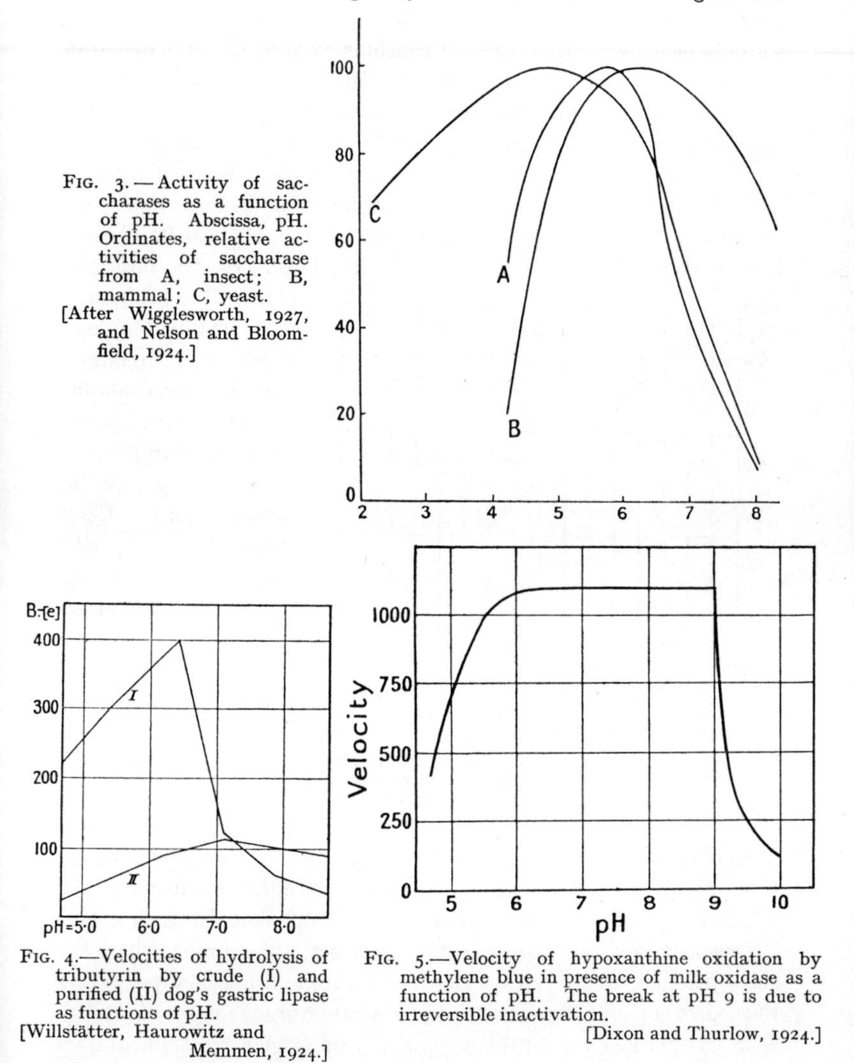

FIG. 3.—Activity of saccharases as a function of pH. Abscissa, pH. Ordinates, relative activities of saccharase from A, insect; B, mammal; C, yeast.
[After Wigglesworth, 1927, and Nelson and Bloomfield, 1924.]

FIG. 4.—Velocities of hydrolysis of tributyrin by crude (I) and purified (II) dog's gastric lipase as functions of pH.
[Willstätter, Haurowitz and Memmen, 1924.]

FIG. 5.—Velocity of hypoxanthine oxidation by methylene blue in presence of milk oxidase as a function of pH. The break at pH 9 is due to irreversible inactivation.
[Dixon and Thurlow, 1924.]

region; or the enzyme may be kept for some time at an unfavourable pH, and its action at the optimal pH subsequently tested. When this is done, and reaction velocities are plotted against pH, curves are obtained such as those of Figs. 2 to 5.

The most usual type is a fairly symmetrical curve such as those of Fig. 3. This type is characteristic of all the enzymes which hydrolyse carbohydrates. Independence of pH over a wide range has only been found in the cases of xanthine-oxidase, bacterial dehydrogenases acting *in situ*, and catalase.[1] The relation between pH and activity may be influenced by a variety of factors. It is often dependent on temperature. Thus in the case of yeast saccharase a rise of temperature pushes the optimum into a more acid region [Nelson and Bloomfield, 1924]. The nature of the buffer is important in some cases, e.g. animal amylases (see Chap. VII), and its concentration often affects the reaction velocity, but rarely the optimum pH. The substrate concentration may affect the relation. This is, for example, the case with honey (bees' salivary) saccharase, as shown by Nelson and Cohn [1924], but not with yeast saccharase on the alkaline side of its optimum. On the acid side, however, the velocity of yeast saccharase action is relatively greater in strong sucrose solutions [Josephson, 1924]. The optimum pH for urease is conspicuously affected by substrate concentration (Fig. 6). Where one enzyme acts on several different substrates the relation between pH and activity may be different for each. This is conspicuously the case with the proteases and erepsins, where the substrates are electrolytes. Hence the optimum pH changes during the course of peptic digestion as the substrate is gradually broken down. It is less markedly so in the case of emulsin (β-glucosidase) acting on different substrates. Where the substrates are not ionized the pH activity relation seems generally to be the same for all substrates.

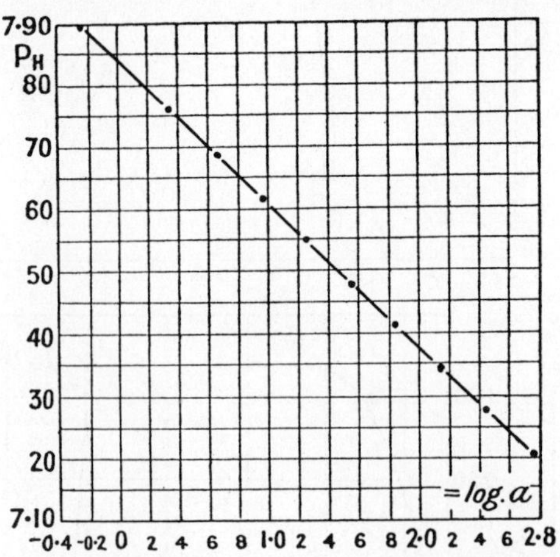

Fig. 6.—pH optimum of Soya urease activity as a function of logarithm of substrate concentration (in mg. per 100 c.c.) [Lövgren, 1921.]

In some cases the pH-activity relation is markedly affected by the presence of organic co-enzymes or activators, for example, in the case of lipase (Fig. 4).

[1] According to Michaelis and Pechstein, but not to Sörensen.

TABLE II.

OPTIMUM pH.

Enzyme.	Source and Condition.	Substrate.	pH.	Reference.
Phosphatase	Mammalian tissues	Monophosphoric esters [1]	9·0-9·2	Kay (1928, 1), Asakawa (1929).
,,	,, ,,	Benzyl and cresyl phosphate	9·4	Asakawa (1929).
,,	,, ,,	Phenyl ,, ,,	abt. 10	,, ,,
,,	,, ,,	Di-esters of phosphoric acid [2]	abt. 7·8	,, ,,
,,	,, ,,	Pyrophosphoric esters	7·5	Kay (1928, 2)
,,	,, ,,	Lecithin	abt. 7·2	,, (1928, 1).
,,	*Helix*	(?)	abt. 3	Karrer (1926).
,,	*Aspergillus* (crude)	Glycerophosphoric acid	5·5	Kobayashi (1927).
,,	,, (purified)	,, ,,	3·6	,, ,,
,,	,,	Propyl and diphenyl phosphates [3]	5·5	Asakawa (1929).
,,	,, (crude)	Diphosphoglyceric acid	5·6	Kobayashi (1929).
,,	,, (purified)	,, ,,	3·0	,, ,,
,,	Malt	Phytin	5·6	Adler (1915).
Lipase	Stomachs of 32 vertebrates	Tributyrin	5·5-8·6	Haurowitz and Petrou (1925).
,,	Man's stomach	,,	6	,, ,, ,,
,,	,, ,, (purified)	,,	8	,, ,, ,,
,,	Pancreas (phosphate buffer)	Ethyl butyrate	7	Platt and Dawson (1925).
,,	,, (borate buffer)	,, ,,	8·5	,, ,, ,,
,,	Liver		8·3	Knaffl-Lenz (1923).
,,	*Pneumococcus*	Tributyrin	abt. 7·5	Avery and Cullen (1920).
,,	*Aspergillus*	,,	8·6	Willstätter and Kumagawa (1925).
,,	*Ricinus*		5	Haley and Lyman (1921).
,,	*Carica papaya*	Olive oil	6·0	Sandberg and Brand (1925).
Pectase	Fruit	Pectin	4·3	Euler and Svanberg (1919).
,,	Tobacco	,,	5·6	Neuberg and Kobel (1927).
Nucleosidase		Adenosine	7·5	Levene, Yamagawa, and Weber (1924).
Maltase	Human gut	Maltose	6·1	Wigglesworth (1927).
,,	Cockroach gut	,,	5·6	,, ,,

[1] Propyl, isopropyl, glycol, α- and β-glyceryl, acetone, phenyl-ethyl, esters, nucleotides, fructose diphosphate.
[2] Ethyl, β-glyceryl, phenyl, cresyl.
[3] Other esters, e.g. benzyl, appear to have an optimum pH near 3.

TABLE II.—*continued.*

Enzyme.	Source and Condition.	Substrate.	pH.	Reference.
Maltase	Yeast	Maltose	6·6	Michaelis and Rona (1913, 4).
,,	,,	α-methyl glucoside	6·2	Rona and Michaelis (1913).
,,	Malt	,, ,,	4·5-5·0	Pringsheim and Leibowitz (1925, 1).
β-glucosidase	Almond	several β-glucosides	4·1-4·5	Josephson (1925, 1).
,,	Salicaceae and *Gossypium*	Salicin	5·2-7·0	Blagoveschenski and Sossiedov (1927).
Cellobiase	*Aspergillus*	Cellobiose	abt. 6	Bertrand and Compton (1911).
,,	Malt	,,	5·1	Pringsheim and Genin (1924).
Melibiase	Almond	Raffinose	4	Willstätter and Csanyi (1921).
Lactase	Cockroach	Lactose	5·8	Wigglesworth (1927).
,,	Yeast	,,	abt. 7	Willstätter and Steibelt (1920).
,,	Almond	,,	4·4	,, and Csanyi (1921).
Saccharase	Mammalian gut	Saccharose	6·2	Wigglesworth (1927).
,,	Bees' saliva [1]	,,	5·5-6·3	Nelson and Cohn (1924).
,,	Cockroach	,,	6·0	Wigglesworth (1927).
,,	*Pneumococcus*	,,	abt. 6·8	Avery and Cullen (1920).
,,	Yeast	,,	4·6-5·0	Nelson and Bloomfield (1924).
,,	*Aspergillus*	,,	5·6	Hattori (1924).
,,	Leaves of 14 angiosperms	,,	4·5-6·6	Blagoveschenski and Sossiedov (1925).
Sinigrinase	Mustard	Sinigrin	abt. 7	Euler and Eriksson (1926).
Amygdalase [2]	Almond	Amygdalin	6·0	Willstätter and Csanyi (1921).
Polyamylase	*Aspergillus*	Polyamyloses	abt. 4·5	Leibowitz and Mechlinski (1926).
Amylase	Salivary and Pancreatic + ? trace of Cl′	Starch	6·0-6·1	Myrbäck (1926, 2).
,,	Salivary and Pancreatic + Cl′, Br′, I′	,,	6·7-6·8	,, ,,
,,	Salivary and Pancreatic + NO_3', ClO_3'	,,	6·9-7·0	,, ,,
,,	Mammalian gut	,,	abt. 7	Rothman, Widener, and Davison (1927).
,,	,, liver	,,	abt. 6	Eadie (1927).
,,	,, blood	,,	abt. 8	Temminck Groll (1924).
,,	Milk	,,	5·9	,, ,, ,,
,,	Ascidian	,,	7·5	Berrill (1929).
,,	Cockroach	,,	abt. 6	Wigglesworth (1927).
,,	Mammalian muscle	Glycogen	abt. 7	Lohmann (1926).
,,	*Pneumococcus*	Starch	abt. 7	Avery and Cullen (1920).

[1] Varies with substrate concentration. [2] Probably mixture of several enzymes.

Enzyme.	Source and Condition.	Substrate.	pH.	Reference.
Amylase	? Bacterial	Starch	7·2	Pringsheim and Schapiro (1926).
,,	Yeast	,,	5·6	Temminck Groll (1924).
,,	*Aspergillus*	,,	4·8	Sherman, Thomas, and Baldwin (1919).
,,	*Cladophora*	,,	4·2	Sjöberg (1921).
,,	5 Phanerogam leaves	,,	5·0-5·4	,, (1922).
,,	Malt	,,	5·2	Myrbäck (1926, 2).
Mannanase	,,	Mannan	abt. 5	Pringsheim and Leibowitz (1925).
Lichenase	,,	Lichenin	abt. 5	,, Seifert (1923).
Inulase	,,	Inulin	3·8	,, Kohn (1924).
Cellulase	Helix hepato-pancreas	Cellulose	5·2	Karrer and Hofmann (1929).
Chitinase	,, ,, ,,	Chitin	5·2	,, ,, ,,
,,	,, ,, ,,	Chitosan	4·5	,, ,, ,,
Urease	*Aspergillus*	Urea	8·6	Bach (1929).
,,	Soya bean	,,	7·2-7·9[1]	Lövgren (1921).
Asparaginase	Yeast	Asparagine	8·0	Geddes and Hunter (1928).
,,	*Aspergillus*	,,	8·5	Bach (1928).
Deaminase	Muscle	Adenylic acid	abt. 6	Schmidt (1929).
Histozym	Mammalian tissues	Hippurate, etc.	abt. 7	Kimura (1929).
Arginase	Liver	Arginine	9·8	Hunter and Dauphinée (1924).
Histidase	,,	Histidine	abt. 9	Edlbacher (1926).
Dipeptidase I	Gut	Simplest[2] dipeptides	7·3	Linderström-Lang and Sato (1929).
,,	Malt	,, ,,	7·8	,, ,, ,, ,,
Dipeptidase II	Gut	,, ,,	8·1	,, ,, ,, ,,
,,	Malt	,, ,,	8·6	,, ,, ,, ,,
,,	Yeast	,, ,,	abt. 8	Willstätter and Grassmann (1926).
Polypeptidase	Gut	,, tripeptides	abt. 7	Waldschmidt-Leitz, Balls, and Waldschmidt-Graser (1929).
,,	Yeast		abt. 7	Grassmann (1927).
Pepsin	Stomach	various proteins	1·5-2·5	Northrop (1922, 5).
Trypsin	Pancreas	,, ,,	8-11	,, ,,
Kathepsin	Spleen, etc.			
Rennin	Stomach	Casein	6·1	Michaelis and Mendelssohn (1914).
,,	*Aspergillus*	,,	abt. 6	Nishikawa (1927).
Protease	Yeast	Gelatin	,, 5	Willstätter and Grassman (1926).

[1] Varies with substrate concentration. [2] I.e. mono-amino monocarboxylic aliphatic peptides.

TABLE II.—*continued.*

Enzyme.	Source and Condition.	Substrate.	pH.	Reference.
Papain	*Carica papaya*	various proteins	5-7·2	Willstätter, Grassman, and Ambros (1926).
Peptidase	14 Angiosperm leaves	Peptone	4·5-8·0	Blagoveschenski and Bielozerski (1925).
Keratinase	Bird's stomach	Keratin	7-8	Stankovic, etc. (1929).
Glyoxalase	Liver	Phenyl-glyoxal	abt. 8	Kuhn and Heckscher (1926).
Carboxylase	Yeast	Keto-acids	4·8	Hägglund and Ringbom (1927).
"Zymase"[1]	Yeast (fructose diphosphate synthesis)	Glucose	6·4	Euler and Nordlund (1921).
,,	Yeast (glucose destruction)	,,	5·5-8	Hägglund and Rosenquist (1926).
,,	,, (CO_2 production)	,,	4·5-6·5	Euler and Heintze (1919).
,,	Lactic bacillus	,,	6·0	Virtanen and Karström (1928).
Aldehyde mutase	Yeast	Aldehyde	abt. 6·5	Myrbäck and Jacobi (1926).
Catalase	Liver	H_2O_2	6·3-9·5	Michaelis and Pechstein (1913).
,,	,,	,,	7	Sörensen (1909).
,,	Blood[4]	,,	7	Nosaka (1927, 1).
Peroxidase	Root	H_2O_2, guaiacol and cresol	abt. 5	Ucko and Bansi (1926).
,,	,,	H_2O_2, HI	abt. 3	Abel (1922).
Succinoxidase	Muscle	Succinate	9·0	Ohlsson (1921).
,, [2]	*Bacillus coli*	,,	8·0-1·00	Quastel and Whetham (1924).
Aldehyde oxidase	Potato	Aldehyde + NO'_3	abt. 5·5	Bernheim (1928, 1).
,, ,,	,,	,, me. blue	abt. 7	,, ,,
Xanthine oxidase	Milk	Xanthine	5·5-8·5	Dixon and Thurlow (1924).
Tyramine ,,	Liver	Tyramine	abt. 10	Hare (1928).
Quinol oxidase	*Lactarius*	Quinol	4·2	Wieland and Sutter (1928).
Laccase[1]	Lac tree	Guaiacol[3]	6·7-8	Fleury (1925).

[1] Probably mixtures of several enzymes. [2] Not in solution. [3] Varies with substrate concentration.

[4] Bodansky (1919) found an optimum of 7·5 if the reaction was allowed to proceed for ten minutes, but of 12 when it was stopped after one minute.

pH *Optima.*

The optimum pH for a number of enzymatic reactions is given in Table II., which summarizes most of the reliable data in the literature. The pH optimum of most if not all lipases depends on accompanying substances (see p. 134) and is changed by purification. The figures given are therefore of only relative significance. That of the carbohydrases is only affected by a few poisons (see p. 150). That of the proteases, peptidases, and phosphatases varies considerably with the substrate, and the same is true of certain oxidases. That of zymase varies according to which of its functions is investigated. For example, in an alkaline solution it produces acetate in place of alcohol. Hence the optimum pH for acetic acid production is far to the alkaline side of that for alcohol. The figures given are of varying reliability. When several workers disagree I have generally given a preference to one of them, which may not always be justified.

The distribution of optima is given in Table III. When an optimum is given at "about 4," or 4·0, I have added ½ to the figures for 3-4, ½ to those for 4-5, and so on. This table does not include figures for lipases, proteases, or certain other enzymes where the optimum varies greatly according to the reaction measured.

TABLE III.

pH range	2-3	3-4	4-5	5-6	6-7	7-8	8-9	9-10	10-11
Number of optima	½	3	19½	27	26	15	8	4·5	½

The range would be slightly extended to the acid side were pepsin included. It will be seen that the median and modal pH optimum is about 6, and the values are distributed round this with a tendency to crowd on the acid side. Some of the groups have a more restricted range. Thus all the carbohydrases (which include the majority of enzymes investigated) have optima between 3·8 and 7·5, all the lipases between 5 and 8. The enzymes acting on bases (urease, asparaginase, arginase, histidase, tyramine oxidase) all have optima more alkaline than pH 7, which probably means that they act on free, as opposed to ionized, base (xanthine oxidase has no definite optimum, pH 7 is in the middle of its range). Other generalizations could no doubt be made.

Michaelis' Theory.

Michaelis and Davidsohn [1911] compared the curves obtained by plotting activity against pH with those obtained when the proportion of molecules of an ampholyte which are uncharged were plotted against pH. In the case of an ampholyte this proportion reaches a maximum at the isoelectric point, and falls off on each side. In the case of a simple ampholyte such as a mono-amino-mono-carboxylic acid, or a peptide composed of several such, the proportion can be expressed in a simple manner. When the pK values of the two ionizations (i.e. the pH's at which ionization is half complete) differ by more than 4 units, the proportion of molecules uncharged is nearly 100 per cent. over a certain range, i.e. the curve obtained by plotting it against pH has a plateau, as in the case of glycine. When the values are nearer together, the proportion of the ampholyte which is unionized never reaches unity, but has a sharp maximum at the isoelectric point, as in the case of *m*-aminobenzoic acid. The pH-activity curves for enzymes mostly resemble these fairly closely. The general equation for these curves is $y = \dfrac{1}{1 + \dfrac{K_1}{cH} + \dfrac{cH}{K_2}}$, where K_1 and K_2 are the two dissociation constants in terms of Brönsted's theory of ampholytes. In the older terminology $K_1 = K_a$, $K_2 = \dfrac{K_w}{K_b}$. When K_1 and K_2 are very unequal, we may write, on the alkaline side of the isoelectric point $y = \dfrac{cH}{K_1 + cH}$, on the acid side $y = \dfrac{K_2}{K_2 + cH}$. The general equation may be expressed as $y = \dfrac{1}{1 + 2\sqrt{\dfrac{K_1}{K_2}} \cosh \mu x}$, where $x = pH - \frac{1}{2}(pK_1 + pK_2)$, $\mu = \log_e 10 = 2{\cdot}30$. In the case of enzymes it is assumed that only the uncharged molecules are active. Thus the optimum coincides with the isoelectric point or zone, and K_1 and K_2 can be calculated for the enzyme.

The behaviour of certain enzymes in an electric field confirmed this view. Thus yeast saccharase does not move when a current is passed through a solution of it near the optimum pH, while it is negatively charged on the acid side of the optimum, positively on the alkaline. The same conclusions as to its charge have been drawn from a study of the action on it of poisonous ions (see Chap. VIII.). In the majority of cases, however, the optimum does not correspond with the isoelectric point of the enzyme preparation. [Michaelis and Davidsohn, 1911, Michaelis and Rona, 1913, Michaelis and Pechstein, 1913]. In these cases it may be assumed that the enzyme is active when it has a certain positive or negative charge, inactive when the charge is too large or too small, or, more probably, that the activity, like the stability of oxyhæmoglobin, depends mainly on the ionization of certain groups in the enzyme molecule. This assumption leads to the Michaelis equation with exactitude, whereas the charge on a colloid varies with pH in a more complicated way. It is perhaps significant that according to Kuhn and Brann [1926] highly purified hæmoglobin shows a maximum peroxidase activity about pH 5·6, its isoelectric point being about 6·8.

Even with yeast saccharase, however, Willstätter, Graser, and Kuhn [1922] showed that purification, which does not alter the relation between its activity and pH, has a slight but significant effect on its charge. It becomes more electropositive on purification. A really accurate comparison of theory and observation has been made by Euler, Josephson, and Myrbäck [1924] for yeast saccharase (Table IV.) on the alkaline side of its optimum. The calculated results are from the formula

$$y = \frac{1}{1 + \frac{3 \times 10^{-7}}{cH}}, \text{ or pH} = 6{\cdot}52 + \log\left(\frac{1}{y} - 1\right).$$

TABLE IV.

pH	5	6	7	8
V. obs.	100	80	26	3·3
V. calc.	100	80	30	5

The observed values are the means of five observations at different sucrose concentrations, and are very accurate. Thus at pH 8, three observations gave 3·3, two 3·4. The same figures were obtained using raffinose as substrate. At pH 8 the observed figures are 30 per cent. below the calculated, and at 8·5 Nelson and Bloomfield [1924] found that the reaction was proceeding at less than half the theoretical rate. The deviations are not due to irreversible inactivation. They may be due to the dissociation of a second hydrogen ion. Any saccharase action by maltase (see p. 100) would influence results in the opposite direction. It is clear therefore that Michaelis' theory is incomplete, though it has been, and still is, a valuable approximation to the facts. In some cases the divergence has been shown (see Chap. III.) to be due to a dependence on pH of the affinity between enzyme and substrate, as well as the rate of breakdown of their compound.

Northrop's Theory.

In the cases of pepsin, trypsin, and papain, it would seem that pH affects the reaction velocity primarily by altering the charge on the substrate rather than the enzyme. Crude pepsin is an ampholyte whose isoelectric point, as judged by cataphoresis, lies in the neighbourhood of pH 2 [Michaelis and Davidsohn, 1910]. Pekelharing and Ringer [1911] found that a purer preparation was always negatively charged. In the case of trypsin, Northrop [1924, 2] found that between pH 2·0 and 10·2 trypsin showed the same distribution coefficient as $\overset{+}{H}$ between particles of gelatin and the surrounding solution, that is to say, it behaved, in accordance with the Donnan equilibrium, as a singly charged cation. In more alkaline solutions (in which it is unstable) it behaved as a singly charged anion.

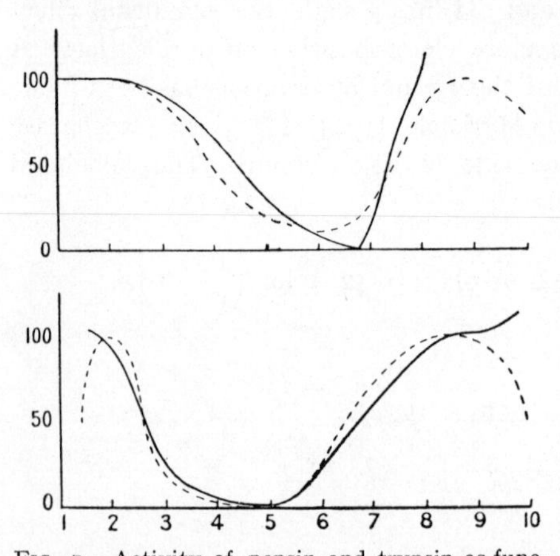

FIG. 7.—Activity of pepsin and trypsin as functions of pH. Upper curves on hæmoglobin, lower on casein. Abscissa, pH. Ordinates, plain lines, titration curves of proteins; dotted lines, rates of digestion (pepsin on left, trypsin on right).

[After Northrop, 1922, 5.]

Northrop [1922, 5] examined the action of pepsin and trypsin on casein, edestin, gelatin, hæmoglobin, and globin. The isoelectric points of these proteins vary from 4·7 (casein) to 6·8 (hæmoglobin). His results for these two proteins are reproduced in Fig. 7. It will be seen that pepsin is only active on a positively charged protein, trypsin on a negatively charged. In the case of pepsin, moreover, there is a fairly good agreement between the reaction velocity and the charge on the protein micelle as determined by conductivity measurements. Since actual measurements show that pepsin unites most readily with its substrate in the neighbourhood of its optimum pH (Fig. 11), there can be little doubt that this union is determined largely by electrical forces. In the case of trypsin the agreement is not quite so good. This is due at least in part to the fact that as pH 10·2 is approached, trypsin begins to lose its charge. In agreement with the above theory Hugounenq and Loiseleur [1925] found that if proteins are diazotized, acetylated, or methylenated [Johanessohn, 1917] so as to alter their free amino groups and destroy their basic properties, they can no longer be digested by pepsin, but can be by trypsin. On the other hand, pepsin either possesses no free amino groups, as suggested by its behaviour as an electrolyte, or if so they do not affect its action, for it is not inhibited by diazotization or by formaldehyde. But both these processes inhibit trypsin, thus supporting the idea that in both cases the enzyme-substrate union is salt-like. As, however, Hugounenq and Loiseleur find that pepsin digests glycyl-glycine, their other results should be accepted with caution. Similarly Imai [1924] found that methylation of proteins, which would have more effect on their properties as acids than as bases, made them more resistant to trypsin, but had little effect on their peptic hydrolysis.

Several plant proteases show an optimum activity in the neighbourhood of the isoelectric point of their substrate. Thus Willstätter, Grassmann, and Ambros [1926], found that papain acted optimally on gelatin, albumin peptone, and fibrin at pH 5, 5, and 7·2 respectively, their isoelectric points being 4·8, 4·8, and 7·2. Yeast protease [Grassmann and Dyckerhoff, 1928, 3], and a number of intracellular proteases from both animals and plants, behave in the same manner (see Chap. VI.).

The same phenomenon is found with the action of erepsin on peptides. Abderhalden and Fodor [1916] found in the case of yeast dipeptidase pH optima varying from 8·5 to 6·8 with different dipeptides. The case of gut erepsin (or rather its dipeptidase components) was attacked by Northrop and Simms [1928]. They found the following optima in ·01M concentration of dipeptide :—

Glycyl-glycine, glycyl-alanine, glycyl-leucine, glycyl-asparagine 7·8-8·0,
Glycyl-aspartic acid 6·0.

No absolute significance attaches to these figures, for the Michaelis constant (enzyme-substrate affinity, see Chap. III.) varies strongly with pH, and hence the pH optimum must vary with the substrate concentration. But the difference between glycyl-aspartic acid and the other dipeptides shows clearly the nature of the substrate's influence on optimal pH. It is difficult to avoid the conclusion that only one ionic species of enzyme unites with only one of substrate, but it is not certain which species are concerned, though it seems likely that an enzyme anion, predominating at pH 7·6, unites with a peptide in the form $R - \overset{+}{N}H_3$ (generally a zwitter-ion). The matter is further complicated by the fact that the optimum is very sensitive to salts, the glycyl-glycine optimum being shifted to pH 6·8 by ·1M $CaCl_2$. This may be due to activation of Linderström-Lang's [1929] dipeptidase I.

The Possible Significance of $\overset{+}{H}$ and $\overset{-}{OH}$ Catalysis.

It is occasionally forgotten that all the hydrolyses and dehydrations catalysed by enzymes, and some of the oxidations and reductions, can also be catalysed by $\overset{+}{H}$ and $\overset{-}{OH}$ ions. The possibility is therefore open that this fact may have a certain relevance for enzyme catalysis. Thus if we suppose that yeast invertase renders sucrose more sensitive to $\overset{+}{H}$ ions, reducing the energy needed for the activation of a hydrated [sucrose + $\overset{+}{H}$] molecule from 26,000 to about 9000 calories per gram-molecule (see p. 67), we must divide the observed velocities by the hydrogen ion concentration to obtain the relation between catalytic efficiency and pH. The curve so obtained has a maximum about pH 7. In this case the maximum catalytic effect would occur, as with many other enzymes, away from the isoelectric point of the enzyme. Moreover, the results of Myrbäck (Chap. VIII.) would be far less intelligible. On the other hand, it is perhaps easier to picture the sucrose molecule, strained by the field of the enzyme in accordance with the theory of Quastel (Chap. X.) as undergoing an exaggeration of its normal sensibility to $\overset{+}{H}$ ions, than as developing a new type of instability. The possibility of a simultaneous union with $\overset{+}{H}$ and $\overset{-}{OH}$ ions may have to be considered in the future.

pH and Enzyme Stability.

Enzymes are generally destroyed by very strong acids or alkalis, but many are inactivated in relatively weak solutions. The phenomenon will be further considered in Chapter IV., as it appears to be the same process as heat-inactivation, regarded from a different point of view. In general, as might be expected, enzymes are fairly stable in the region of their optimum pH. But this is by no means invariably the case. Thus trypsin (Fig. 8) is most stable at pH 6, or in presence of inhibitor (see Chap. III.) at pH 5, the optimum for digestion being pH 8-10. Similarly pepsin is most stable at pH 4

(Fig. 9). It is clear from Ege's results that in this case two different processes are at work, namely, destruction by $\overset{+}{H}$ and $\overset{-}{OH}$ ions. On

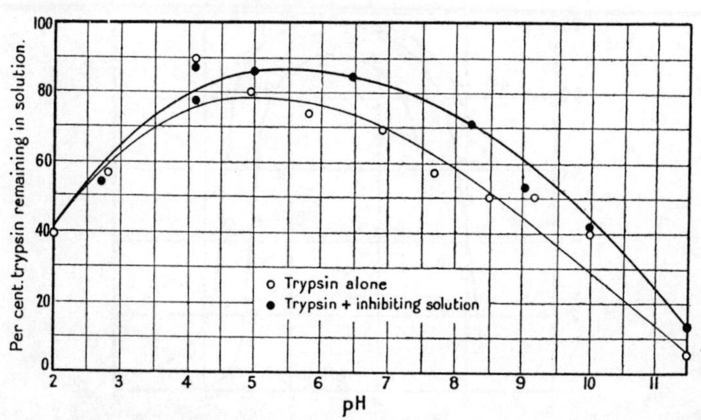

FIG. 8.—Percentage of trypsin remaining after 30 minutes at 38° C. as a function of pH. [Northrop, 1922, 3.]

the other hand, Northrop [1920, 1], working with a different preparation, found much less effect of pH in acid solutions.

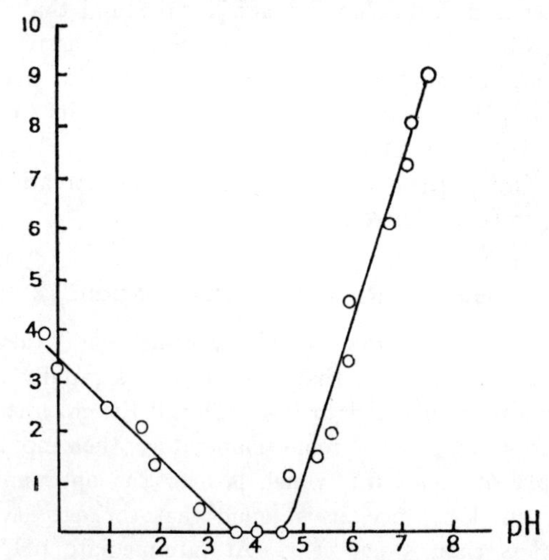

FIG. 9.—Logarithm of velocity of pepsin destruction as function of pH (temperatures, 10° C. to 60° C., but all observations reduced to same temperature). [Ege, 1925.]

Yeast saccharase, on the other hand, is most stable near its optimum pH (Fig. 10). Malt amylase, according to Ernström [1922], has an optimum stability in acetate buffer at pH 5·5, in phosphate at pH 6, whereas the optimum for reaction velocity is about pH 5.

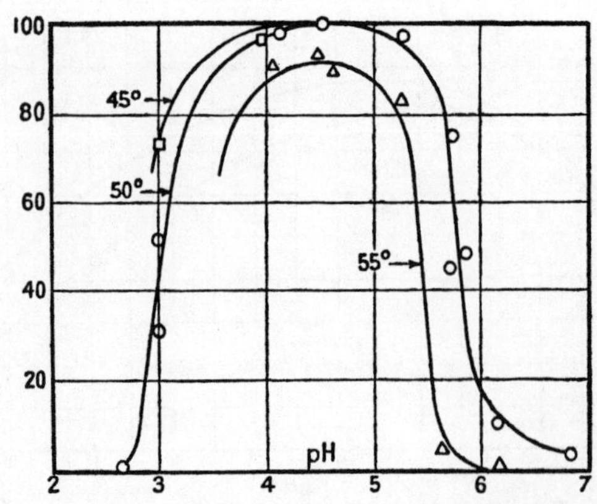

FIG. 10.—Percentage of yeast saccharase remaining after 1 hour at 45° ⊡, at 50° O, at 55° △, as a function of pH. [Euler and Laurin, 1919.]

Activation by Acids.

Willstätter and Waldschmidt-Leitz [1924] found that after treatment with HCl, castor-oil bean lipase, previously inactive at pH 6·8, and having an optimum at pH 5·0, became active at pH 6·8, while its activity at pH 4·7 was reduced to one-third of its former value, and was less than that at pH 6·8. The process is perhaps to be regarded as a shift in the optimum rather than an activation. It may be due to hydrolysis of a co-enzyme.

Slowly Reversible Inactivation.

Ohlsson and Swaetichin [1929], working on "takadiastase" (*Aspergillus* amylase), found that this enzyme is rapidly inactivated at a pH which is unduly high or low. Thus if the enzyme is kept for ten minutes to pH 2 or 12 at room temperature, then rapidly brought back to a pH of about 6·1, which is near the optimum, and the activity measured at once, it is found that the activity has been reduced to less than 5 per cent. At intermediate pH values inactivation is slower, thus at pH 3·2 half the activity is lost in 100 minutes. If, however, the enzyme is left at pH 6·1 for an hour, a great deal of the initial activity is regained. Thus an enzyme which had been reduced to 4 per cent. activity by ten minutes' exposure to pH 2·05, returned to 60 per cent. activity after one hour at pH 6·1.

In twenty-one hours it had only risen to 63 per cent. The speed and extent of reactivation depend on pH. The optimum pH for reactivation is about 6·5, whereas that for enzymic activity is about 5·5. However, although the final activity reached is highest at pH 6·5, the velocity of reactivation is greater at higher pH values. Not only inactivation, but reactivation, are slowed down by the presence of starch. Hence if enzyme is added to starch at pH 6·1 after passing some time at pH 1, there is very little hydrolysis during the first hour, and the velocity of hydrolysis gradually increases during the first part of the reaction, instead of remaining constant or falling slightly, as with normal enzyme.

It is clear from these experiments that the reversibility or otherwise of pH effects depends on details of technique, and inactivation will appear to be reversible with one technique, irreversible with another.

CHAPTER III.

THE UNION OF ENZYME WITH ITS SUBSTRATE AND RELATED COMPOUNDS.

WITH the exception of Barendrecht [1924], who supported a radiation theory of enzyme action, almost all authors have postulated a union of enzyme and substrate. But while some have favoured a chemical union, others have postulated adsorption on the surface of the enzyme particles or molecules. While recent work in physical chemistry, notably that of Langmuir [1916], has tended to obliterate the sharpness of such a distinction, two points may at once be raised with regard to adsorption. In the first place, since enzyme molecules are probably of the same order of magnitude as those of proteins and starch, only a small number at most of substrate molecules can be united with one of protease or amylase. And secondly, the number is not very large even where the substrate is simpler.

Thus an enzyme molecule of molecular weight 60,000, and density 1·1, would have a radius of 28 Ångstrom units if spherical, and an area of 9856 square Ångstroms. The area occupied by a triglyceride in a condensed film at a water-air interface is 63·0 square Ångstroms. Such an enzyme molecule could therefore accommodate 156 fat molecules, if they were packed as closely as possible. This figure would be increased four times if the enzyme molecule were eight times as large, and so on. It would also be increased if the enzyme molecules were not spherical. Thus, if the molecule were a sheet 10 Å. thick, its flat surfaces would have an area of 18,000 Å., and it could accommodate about twice as many fat molecules. The number of molecules of a biose or hexoside, since there is reason to think that the whole ring of a hexose molecule is in contact with the enzyme, would probably be about half this. But it is certain that the whole enzyme surface is not catalytically active, for Michaelis [1908] and Nelson and Griffin [1916] found that yeast saccharase loses none of its activity when adsorbed by a variety of solids, and several, though not all, other enzymes behave in the same way. If the catalytic activity is confined to a few spots on the surface, as seems likely, we may legitimately, at least as a first approximation, assume that the events at each go on independently. This assumption has worked fairly well in the case of hæmoglobin, of which, according to Adair [1928] each molecule can take up four molecules of O_2. If as Willstätter [1922] believes, an enzyme consists of a small chemically active group and a large colloidal carrier, the active centres must be very few in number. In many cases, as we shall see, the assumption that an enzyme molecule activates only one substrate molecule at a time has worked remarkably well. But provided the different active points are far enough away to be independent, the same equations are reached in the case of a large active surface, for example, a platinum surface catalysing gaseous oxidations [Langmuir, 1916].

The Adsorption of Enzyme by its Substrate.

Where an enzyme acts on a solid substrate it is generally adsorbed by it. This proves little, as enzymes are adsorbed by a variety of solids. Nevertheless, in some cases the adsorption is specific. Thus Willstätter and Waldschmidt-Leitz [1922] found tristearin a very effective adsorbent for pancreatic lipase. Moreover, lipase is partially active while adsorbed on kaolin or alumina, but quite inactive when adsorbed on fat or cholesterol. This suggests that it is adsorbed to the latter by its catalytically active group. Similarly casein and fibrin adsorb activated but not inactive trypsinogen [Waldschmidt-Leitz, Schäffner and Grassmann, 1926]. Fibrin also adsorbs enterokinase. On the other hand, this specificity is often absent. Thus Nishikawa [1927] found that fibrin adsorbed amylase, lipase, and rennin, as well as trypsin, from "takadiastase." Northrop [1919] found a maximum removal of pepsin from solution by boiled egg albumin at pH 2 (Fig. 11) which is also the optimum for digestion over short periods, according to Sörensen [1909]. At a later stage in digestion the optimum changes towards pH 1. This agreement suggests that the main effect of pH on peptic digestion is to favour the union of enzyme and substrate. Equally suggestive are the experiments where the enzyme is removed from an adsorbent by a solution of its substrate. In some cases this removal is specific. Thus Michaelis [1921] found that yeast invertase is not removed from colloidal ferric hydroxide by water, or solutions of fructose, lactose, and several other sugars, but is so by its substrates sucrose and raffinose, and also by maltose. Willstätter and Kuhn [1921], however, found that sucrose was the only sugar capable of removing it from alumina. These effects might have been due to the sugar competing for the adsorbent with the enzyme. Hedin [1907] found that casein removed trypsin from adsorption by charcoal, and that the amounts removed increased with the total amount of casein in the system, but were independent of the volume of water in which it was dissolved. It is therefore clear that casein and charcoal were competing for the trypsin, not casein and trypsin for the charcoal. In fact trypsin must unite with dissolved casein.

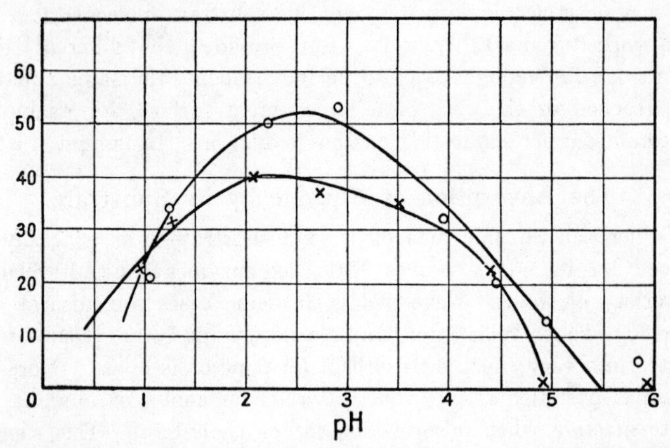

FIG. 11.—Percentages of pepsin removed from solution by coagulated egg albumin in two experiments, as functions of pH. [Northrop, 1919.]

The Effect of Varying Substrate Concentration on Reaction Velocity.

If the concentration of substrate be varied, the initial velocity of reaction at first increases, then reaches a maximum, and finally may diminish. The velocity is at first generally nearly proportional to the substrate concentration (Fig. 12). In many cases there is a fairly wide range of substrate concentrations over which the velocity is constant (Figs. 13, 14), and the falling off in strong solutions is only serious at concentrations where deviations from the mass-action law may be expected to be serious.

The best data for concentrated solutions are those for yeast saccharase. Nelson and Schubert [1928] found a maximum velocity of sucrose hydrolysis at about 5 per cent., and after this a falling off, the relation between sucrose content and velocity being nearly linear between 10 per cent. and 70 per cent. sucrose. In order to determine whether this was due to a falling off in water content they compared the results when the water content was lowered by the addition of sucrose and of alcohol. The differences found were small (Fig. 15), and may be explained on the assumption that the

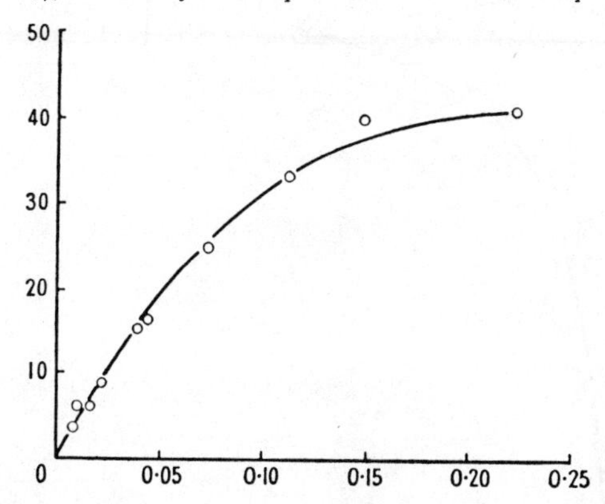

Fig. 12.—Rate of raffinose hydrolysis by yeast invertase as a function of substrate concentration. Abscissa, molar concentration of raffinose. Ordinate, initial velocity of hydrolysis. [After Kuhn, 1923, 1.]

sucrose and alcohol were partly hydrated, and hence the amount of free water was somewhat different to that calculated. Water is, of course, a reactant in the reaction catalysed, and it is clear that

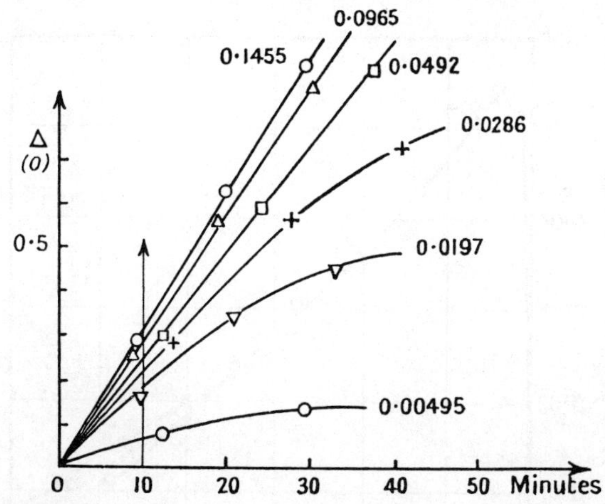

Fig. 13.—Course of hydrolysis of saccharose by yeast saccharase at different initial substrate concentrations. [Kuhn, 1923, 1.]

over a wide range the velocity is proportional to the water concentration. The affinity of the enzyme for water, if any, must be very low, K_m exceeding 100. The inhibitory effects of methyl and ethyl alcohols

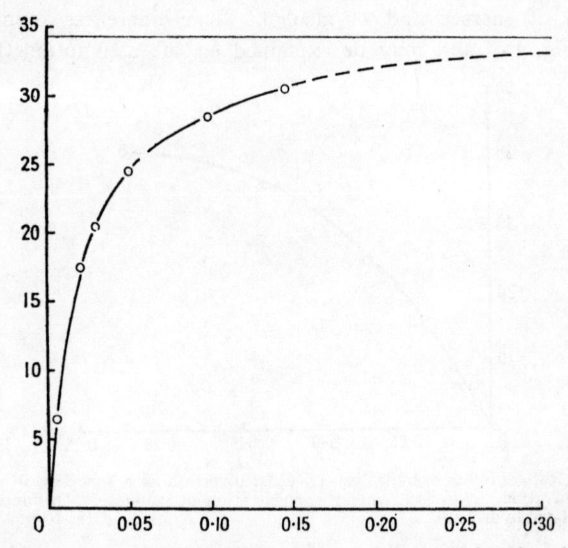

FIG. 14.—Rate of sucrose hydrolysis by yeast saccharase as a function of substrate concentration (from data of Fig. 13). Abscissa, molar concentration of sucrose. Ordinate, initial velocity of hydrolysis. [After Kuhn, 1923, 1.]

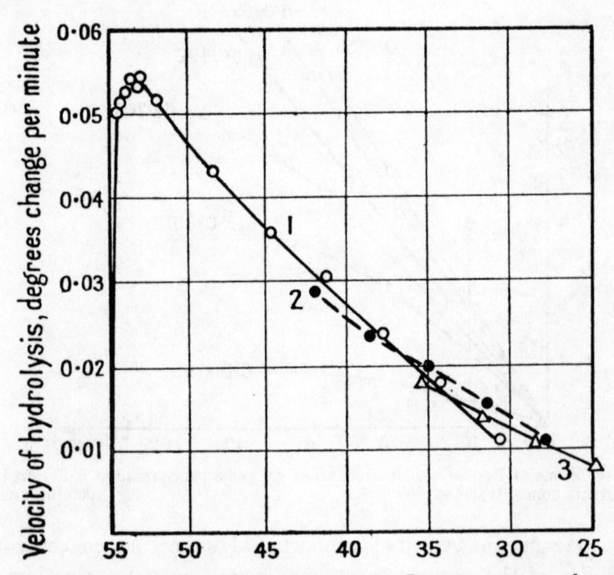

FIG. 15.—Concentration of water, moles per litre. Curve 1, sucrose and no alcohol; 2, sucrose and 10 per cent. alcohol; 3, sucrose and 20 per cent. alcohol. [Nelson and Schubert, 1928.]

on the initial rate of β-glucoside hydrolysis by prunase [Josephson, 1925, 1] are mainly, but not wholly, explicable as due to dilution of the water.

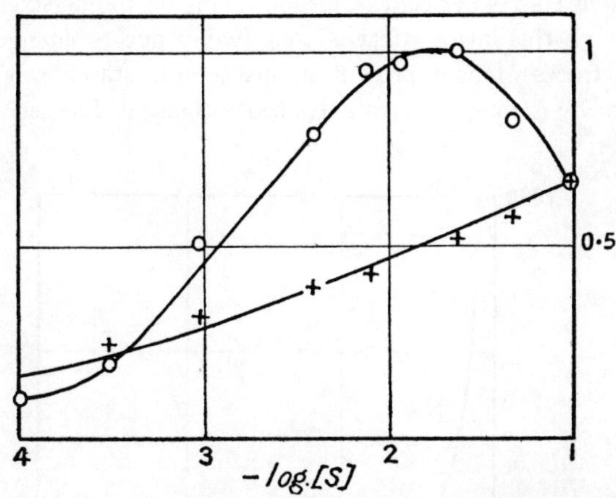

FIG. 16.—Velocities of hydrolysis of ethyl *d*-mandelate (+) and ethyl *l*-mandelate (○) by dog's liver lipase as functions of the logarithm of substrate concentration. [Bamann, 1929.]

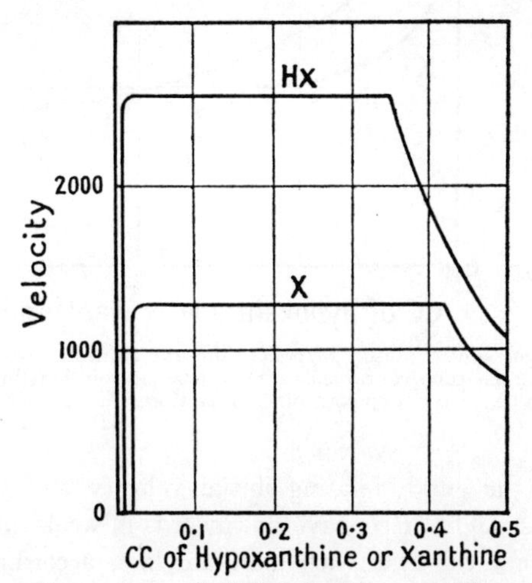

FIG. 17.—Milk xanthine-oxidase. Initial part of hypoxanthine and xanthine concentration curves. Each tube contained: 5·0 c.c. strong enzyme solution (in buffer); 0·5 c.c. methylene blue; *x* c.c. hypoxanthine or xanthine (2 mg. per c.c.); (0·5 – *x*) c.c. water. [Dixon and Thurlow, 1924.]

No explanation of this type is possible in the case of urease (Fig. 31), where the inhibitory effect of strong substrate concentrations is combated by glycine, or liver lipase (Fig. 16); in those of xanthine-oxidase

(Figs. 17, 18) peroxidase, and citric oxidase [Bernheim, 1928] where the effect is very marked in quite weak solutions, or in that of salivary amylase (ptyalin) where Lovatt Evans [1912] found a maximum hydrolysis in 1·5-2·5 per cent. solution, falling off to one-sixth in 5 per cent. (though this latter effect is conceivably due to changes of pH, owing to traces of acid present in his soluble starch preparation). Inhibition by strong substrate concentrations is further discussed in Chapter V.

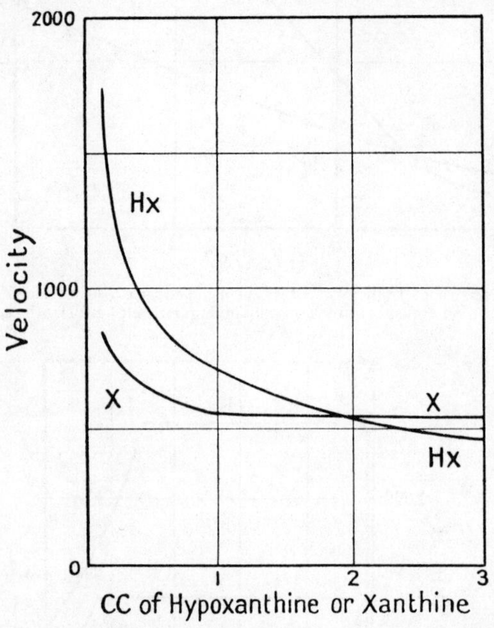

FIG. 18.—Milk xanthine oxidase. Hypoxanthine and xanthine concentration curves compared. Each tube contained: 4·0 c.c. enzyme solution (in buffer); 0·5 c.c. methylene blue; x c.c. hypoxanthine or xanthine (2 mg. per c.c); $(3 - x)$ c.c. water. [Dixon and Thurlow, 1924.]

Generally the curve relating initial velocity and substrate concentration is straight or concave downwards in weak solutions. This is, however, not the case with malt amylase, according to Sjöberg and Ericcson [1924], confirmed by Eadie [1926], where no action whatever occurs in sufficiently weak concentrations of amylose, amylopectin, or glycogen (see Fig. 20). It is also quite possible (see p. 52) that the initial position of the curve for insect saccharase is concave upwards.

Table V. gives a list of the substrate concentrations at which half

TABLE V.

SUBSTRATE CONCENTRATIONS GIVING HALF MAXIMUM VELOCITY (IN MANY CASES IDENTICAL WITH MICHAELIS' CONSTANT).

Enzyme.	Source.	Substrate.	Concentration.	Reference.
Phosphatase	Bone	Glycerophosphate	<·003 M	Martland and Robison (1927).
,,	*Aspergillus*	,,	·09 M	Kobayashi (1926).
Lipase	Pig pancreas	Ethyl butyrate	> ·03 M	Willstätter and Memmen (1924, 1).
,,	,, ,,	Methyl ,,	> ·09 M	,, ,, ,,
,,	,, ,,	Methyl *d*-mandelate	·016 M	Weber and Ammon (1929).
,,	,, ,,	,, *l*-mandelate	·016 M	,, ,, ,,
,,	,, liver	,, *d*-mandelate	·00018 M	,, ,, ,,
,,	,, ,,	,, *l*-mandelate	·0013 M	,, ,, ,,
,,	,, ,,	Ethyl *d*-mandelate	·0005 M	Willstätter, Kuhn and Bamann (1928).
,,	,, ,,	,, *l*-mandelate	·0016 M	,, ,, ,, ,,
,,	Sheep liver	,, *d*-mandelate	·0007 M	Bamann (1929).
,,	,, ,,	,, *l*-mandelate	·0017 M	,, ,,
,,	Rabbit liver	,, *d*-mandelate	·0021 M	,, ,,
,,	,, ,,	,, *l*-mandelate	> ·01 M	,, ,,
,,	Dog liver	,, *d*-mandelate	·0011 M	,, ,,
,,	,, ,,	,, *l*-mandelate	> ·002 M	,, ,,
,,	Man's pancreas	,, *d*-mandelate	·0017 M	,, ,,
,,	,, ,,	,, *l*-mandelate	> ·01 M	,, ,,
,,	Ox pancreas	,, *d*-mandelate	> ·02 M	,, ,,
,,	,, ,,	,, *l*-mandelate	> ·05 M	,, ,,
,,	Horse pancreas	,, *d*-mandelate	> ·02 M	,, ,,
,,	,, ,,	,, *l*-mandelate	> ·05 M	,, ,,
,,	Sheep liver	Methyl butyrate	·0013 M	Bamann and Schmeller (1929).
,,	Rabbit liver	,, ,,	·0010 M	,, ,, ,,
,,	Dog liver	,, ,,	·0028 M	,, ,, ,,
,,	Man's liver	,, ,,	·008-·009 M	,, ,, ,,
,,	Ox liver	,, ,,	> ·02 M	,, ,, ,,
,,	Horse liver	,, ,,	> ·03 M	,, ,, ,,
Nucleosidase	Liver	Adenosine	> ·02 M	Levene, Weber and Yamagawa (1924).
Maltase	Yeast	Maltose	·12-·30 M	Willstätter, Kuhn, and Sobotka (1924).

TABLE V.—*continued.*

Enzyme.	Source.	Substrate.	Concentration.	Reference.
Maltase	Yeast	α-methyl glucoside	·037-·075 M	Willstätter, Kuhn, and Sobotka (1924).
,,	,,	α-phenyl glucoside	·021-·050 M	,, ,, ,, ,,
Prunase (=β-glucosidase)	Almond	β-glucose	·185 M	Josephson (1925).
,, ,,	,,	Helicin	·016 M	,, ,,
,, ,,	,,	Arbutin	·042 M	,, ,,
,, ,,	,,	Salicin	·017-·035 M	Willstätter, Kuhn, and Sobotka (1923).
,, ,,	,,	β-methyl glucoside	·60-1·12 M	,, ,, ,, ,,
,, ,,	,,	β-phenyl glucoside	·040-·065 M	,, ,, ,, ,,
Saccharase	Yeast	Saccharose	·016-·04 M	Kuhn (1923, 1).
,,	,,	Raffinose	·24-·66 M	,, ,,
,,	Honey	Saccharose	·02 M	Nelson and Cohn (1924).
,,	Gut	,,	·02 M	Cajori (1930).
Lactase	Almond	Lactose	abt. ·1 M	Armstrong (1924).
" Emulsin "	,,	Amygdalin	·003 M	Auld (1908).
Amylase	Saliva	Starch	0·4 per cent.	Lovatt Evans (1912).
,,	,,	,,	0·08 ,,	Myrbäck (1926, 2).
,,	Pancreas	,,	0·25 ,,	Kendall and Sherman (1910).
,,	Liver	,,	·095 ,,	Eadie (1927).
,,	Muscle	Glycogen	> 0·5 ,,	Lohmann (1926).
,,	Malt	Amylose	0·5 ,,	Sjöberg and Ericsson (1924).
,,	,,	Amylopectin	0·4 ,,	,, ,, ,,
,,	,,	Glycogen	abt. 1 ,,	Eadie (1926).
Urease	Soya bean	Urea	abt. ·025 M[1]	Van Slyke and Cullen (1914), etc.
Asparaginase	*Aspergillus*	Asparagine	< ·05 M	Bach (1928).
Erepsin	Gut	Glycyl-glycine	< ·05 M	Euler and Josephson (1926).
,,	,,	Glycyl-leucine	·02-·07 M[1]	Northrop and Simms (1928).
Trypsin	Pancreas	Casein (conductivity)	2 per cent.	Bayliss (1904).
,,	,,	Casein (amino N)	·75 ,,	Northrop (1922, 4).
,,	,,	,, (first stage)	< 5 ,,	,, ,,
,,	,,	Gelatin (amino N)	·5 ,,	,, ,,
Pepsin	Stomach	Ovalbumin	4·5 ,,	,, (1920, 3).
Glyoxalase	Liver	Methyl-glyoxal	<·005 M	Kuhn and Heckscher (1926).

[1] Varies with pH.

Enzyme.	Source.	Substrate.	Concentration.	Reference.
Glyoxalase	Liver	Phenyl-glyoxal	> ·01 M	Kuhn and Heckscher (1926).
Fumarase [1]	*B. coli*	Fumaric acid	< ·04 M	Woolf (1929).
Carboxylase	Yeast	Pyruvic acid	·01 M	Hägglund and Rosenquist (1927).
Zymase [1]	,,	Glucose	·006 M	Euler and Myrbäck (1923, 2).
Myozymase	Muscle	Amylose	0·1 per cent.	Meyerhof (1926).
Catalase	Liver	H_2O_2	·025 M	Euler and Josephson (1927, 1).
,,	Yeast	,,	> ·02 M	Issajew (1904).
Hæmin (as catalase)	Blood	,,	·008 M	Euler and Josephson (1927, 3).
Peroxidase	Root	H_2O_2, leucomalachite green	6×10^{-6} M	Willstätter and Weber (1926, 1).
Mean tissue oxygenase [1]	*Tenebrio*	O_2	$< 1{\cdot}5 \times 10^{-6}$ M	Thunberg (1905).
Mean tissue oxygenase [1]	Yeast	,,	$< {\cdot}5 \times 10^{-7}$ M	Warburg (1927).
Mean tissue oxygenase [1]	*Micrococcus*	,,	$< 10^{-8}$ M	Warburg and Kubowitz (1929).
Lactic oxidase [1]	Muscle	Lactate	·007 M	Ahlgren (1925).
,, ,,	Yeast	,,	abt. ·005 M	Bernheim (1928, 2).
,, ,,	,,	α-Hydroxybutyrate	abt. ·015 M	,, ,,
Succinoxidase [1]	Muscle	Succinate	·001 M	Widmark (1922).
Malic oxidase	Cucumber	Malate	·007 M	Thunberg (1929).
Citric oxidase	Liver	Citrate		Bernheim (1928).
,, ,, [1]	Muscle	,,	·0005 M	Ahlgren (1925).
,, ,,	Cucumber	,,	2×10^{-5} M	Thunberg (1929).
Glutaminic oxidase [1]	Muscle	Glutaminate	·0005 M	Ahlgren (1925).
Xanthine oxidase	Milk	Adenine, Xanthine, Hypoxanthine	$< 3 \times 10^{-5}$ M	Dixon and Thurlow (1924).
,, ,,	,,	Acetaldehyde	> 1 M	,, ,, ,,

[1] Not in solution.

the maximum initial velocity is reached for a variety of enzymes. These values are on the whole much less certain than the pH optima. They may, moreover, differ for samples of the same enzyme from different sources or differently treated, when the pH effects do not. In the case of the proteases they differ according to what is measured. Thus Northrop [1922, 4] found that the rate of casein digestion, as measured by increase of amino N, was unaltered when the concentration of casein was increased from 3 to 5 per cent., whereas the amount rendered incapable of precipitation by trichloracetic acid was almost exactly proportional to the concentration between 1·5 per cent. and 5 per cent. Presumably casein has a less affinity for trypsin than have some of the products of its digestion.

It will be noticed that a number of the oxidizing-reducing enzymes are saturated by their substrates at considerably lower concentrations than any of the hydrolytic enzymes acting on crystalloidal substrates. Aldehydes with xanthine-oxidase form an exception, but xanthine-oxidase has a very high affinity for its specific substrates. The reducing substrates of peroxidase appear to act, not by combining with it as a preliminary to oxidation, but by combating the inhibition of its activity by the relatively high H_2O_2 concentrations used (see p. 59). The figures obtained for them are therefore not included in the table. Catalase occupies an intermediate position as regards affinity. This may well prove to be a fundamental distinction between hydrolytic and oxidizing-reducing enzymes. Many of the latter also agree in giving a falling off in activity when the substrate concentration is increased beyond the optimum. This is still true even when full allowance has been made for the destructive effect on peroxidase of H_2O_2. It may be that this phenomenon is quite a general one, which, however, in the case of many hydrolytic enzymes cannot be distinguished from the effects of mere diminution of the amount of water in solution.

Michaelis' Theory.

Michaelis and Menten [1913] developed a theory adumbrated by Brown [1902], Henri [1903] and others, and applied it to the initial velocities of hydrolysis of sucrose by yeast invertase in varying sucrose concentrations. If E represent an enzyme molecule, S, G, and F molecules of sucrose, glucose, and fructose, they assume the following reactions to occur :—

$$1.\ E + S \rightleftharpoons ES.$$
$$2.\ ES + H_2O \rightarrow E + G + F.$$

Now let e be the total molar concentration of enzyme, x of sucrose (supposed to be very much greater than e), and p the concentration of ES molecules, so that the concentration of E molecules is $e - p$.

Then if K_m be the dissociation constant of the compound ES,

$$\therefore K_m p = x(e - p), \quad \therefore p = \frac{ex}{K_m + x},$$

and if k be the velocity constant of the break-up of ES, and v the velocity $\frac{-dx}{dt}$ of hydrolysis, then the amount of H_2O is nearly constant,

$$\therefore v = kp$$

$$= \frac{kex}{K_m + x},$$

or if V be the velocity when x is large compared with K_m,

$$v = \frac{Vx}{K_m + x}, \text{ or } K_m = x\left(\frac{V}{v} - 1\right).$$

Plotted graphically the above equation gives a rectangular hyperbola (Fig. 14). K_m is the substrate concentration at which half the limiting velocity is reached.

We have thus two constants available for fitting a series of results, but V should be (and is found to be) proportional to the enzyme concentration, while K_m, which is generally called the Michaelis constant, is a characteristic of the enzyme. The above equation for the velocity is followed, probably within the limits of experimental error, by saccharase, provided the sucrose concentration is not too high, in which case H_2O concentration becomes a limiting factor. Thus in Michaelis and Menten's experiments the velocity, in degrees of polarimetric reading per minute, agrees well with the values calculated from V = ·0786, K_m = ·0167.

TABLE VI.

Molar concn. sucrose	·7700	·3850	·1920	·0960	·0480	·0308	·0154	·077
v. obs	·063	·075	·075	·0682	·0583	·0500	·0350	·0267
v. calc. . . .	·077	·075	·072	·0670	·0583	·0507	·0382	·0244

On the alkaline side of the optimum, K_m does not vary appreciably down to pH 8, but on the acid side it increases, being about doubled at pH 2·7, according to Josephson [1924]. Where K_m does not vary with the pH, it follows that the pH-activity curve is independent of the substrate concentration. Hence saccharase and its sucrose compound have the same acid dissociation constant, i.e. lose $\overset{+}{H}$ equally readily. But on the acid side the free enzyme is a stronger base (or weaker acid) than the compound, as if the sucrose combined with an amino group of the enzyme. Nelson and Anderson [1926] found K_m independent of temperature, whilst Euler and Laurin [1920] with a different saccharase preparation found a diminution of 34 per cent. on raising the temperature from 1° to 39°, giving a heat of reaction of only 2000 calories per gram molecule. This has a considerable bearing on the nature of the union.

A Possible Criticism of the Above Theory.

It has been assumed above that the combination of enzyme and substrate is always in equilibrium, i.e. that the velocities of formation and dissociation of their compound are infinite in comparison with that of its decomposition to form the products of the reaction. This is rather an improbable assumption, for it seems likely that an invertase molecule can invert about 2000 sucrose molecules per second at 15° C., so that the half-period of the reaction $ES \rightarrow E + G + F$ is less than ·0005 second. It seems a little rash to postulate half-periods of 10^{-5} second or less for the other reactions. For only 10^7 cane-sugar molecules in a 0·1 N solution collide with a given point on the enzyme surface per second, and we cannot suppose that the orientation would always permit of union with the enzyme.

Hence Briggs and Haldane [1925] were led to consider the situation when the velocities of the three reactions are comparable.

Let the velocity constants of the reactions

$$E + S \rightarrow ES,\ ES \rightarrow E + S, \text{ and } ES \rightarrow E + G + F$$

be k_1, k_2, k_3 respectively, e and p having their previous significance.

Then $$\frac{dp}{dt} = k_1x(e - p) - k_2p - k_3p.$$

But so long as the velocity of the reaction is constant, p is constant, and even when it is altering, the rate of change of p must be infinitesimal compared with that of x. Hence $\frac{dp}{dt}$ may be taken as zero,

$$\therefore k_1x(e - p) = (k_2 + k_3)p$$

$$\therefore p = \frac{k_1ex}{k_1x + k_2 + k_3}$$

$$= \frac{ex}{x + \dfrac{k_2 + k_3}{k_1}}.$$

Hence provided we take $\frac{k_2 + k_3}{k_1} = K_m$, the result is the same as that of Michaelis and Menten. Van Slyke and Cullen [1914] arrived at the same equation for urease, on the assumption that both reactions are irreversible, i.e. $k_2 = 0$. Actually then we cannot, with methods at present available, estimate the magnitude of k_2, i.e. the velocity of the reaction $ES \rightarrow E + S$. But the importance of the Michaelis constant is undiminished, though it need not be a dissociation constant. It has, however, the dimensions of such a constant, and should be expressed as a molar or percentage concentration.

In spite of this criticism it would seem that Michaelis and Menten's theory must be substantially correct for yeast invertase between pH 4 and 8. Over this range K_m does not alter with pH, whilst k_3

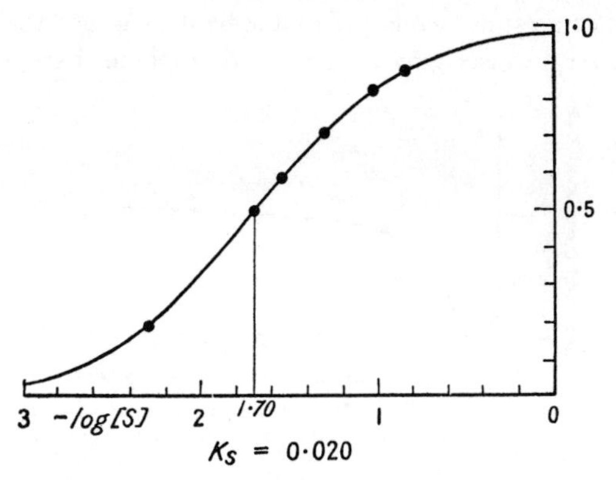

FIG. 19.—Velocities of initial saccharose hydrolysis as a function of the logarithm of substrate concentration, from data of Fig. 13. K_m = ·020. [Kuhn, 1923, 1.]

varies twenty-fold. Unless then we make the quite improbable assumption that both k_1 and k_2 vary proportionally with it, the constancy of K_m shows that k_2 is large compared with k_3. On the other hand, in the case of such enzymes as urease, where K_m varies considerably with pH, either the theory of Van Slyke and Cullen, or that of Briggs and Haldane may turn out to be applicable. The question is further considered in Chapter V., pages 80-83.

In order to calculate K_m from a given set of data, it is convenient to plot the velocity v against the logarithm of the concentration x. Since $\log x = \log K_m - \log \left(\frac{V}{v} - 1\right)$, the curve obtained (Fig. 19) is of the same form as when the amount of dissociation of a weak electrolyte is plotted against pH. The logarithm of the substrate concentration is usually called pS. The advantage of this method is that the points corresponding to low concentrations, which are important in determining K_m, are not crowded together, as would otherwise be the case. The affinity, which is the reciprocal of K_m, is often denoted by the symbol K_M.

Exceptions to the Michaelis Law.

In the case of the oxidases K_m (if the expression has a meaning) is often so small that the relation between v and x in sub-optimal substrate concentrations is still unknown. Liver catalase, and hæmin acting as a catalase, obey the Michaelis law within the limits of experimental error [Euler and Josephson, 1927, 2, 3], K_m being three times as great for the enzyme, although it is at least 10,000 times as active a catalyst as hæmin per unit weight. Among the hydrolytic enzymes several causes may contribute to apparent exceptions. The

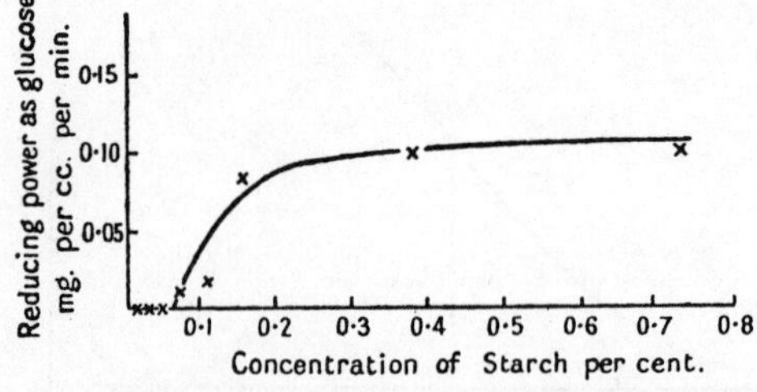

FIG. 20.—Initial velocities of starch hydrolysis (measured by maltose production by malt amylase, as a function of starch concentration in per cent. [Eadie, 1926.]

proteases present great difficulties in the measurement of reaction velocity as they catalyse a number of different reactions. In other cases the enzyme may be inactivated by its substrate or impurities in it. Or the " enzyme " may consist of several components, and the velocity-substrate relation depend on their ratio, as is probably the case with urease.

Yet in many cases it is hard to escape the conclusion that the Michaelis law does not hold. This is conspicuously so where no action takes place below a certain limiting velocity (Fig. 20), but in a few cases the deviation consists largely in a failure of the v, x curve to reach an asymptote, and it is more nearly represented by a " parabola " of equation $v = ax^b$ (where b is less than 1) than by a rectangular hyperbola. Just as the hyperbola is the isotherm of a monomolecular reaction, so is such a curve that of an adsorption isotherm according to Freundlich's [1922] equation. Nevertheless, such a curve has not yet been found to fit a v, x curve as accurately as the rectangular hyperbola of Michaelis fits it in favourable cases. In such cases defenders of the Michaelis equation might perhaps suggest that the enzyme consists of a mixture of different chemical individuals, each with its own K_m, or that the several reactive centres on a single enzyme molecule have somewhat different values of K_m. This is plausible, because K_m is much less specific than optimum pH. It varies considerably with invertase preparations from different yeasts, and even with different preparations from the same yeast, whilst their pH behaviour is always the same. The velocity equation would then be of the form

$$v = \frac{V_1 x}{x + K_1} + \frac{V_2 x}{x + K_2} + \ldots$$

which would allow of a relatively slow approach to the asymptote.

In a few cases the asymptote is approached too rapidly for the Michaelis equation, as in the case of taka-invertase where the relation between velocity and substrate concentration remains almost linear until saturation is reached, according to Hattori [1924]. This is equally hard to explain on the combination or adsorption theories. In Eadie's [1926] case a very good fit of the observed relation was obtained with the equation $v = a \log x - b$, which Quastel and Whetham [1925, 1] had successfully applied to surface catalyses where no action occurs below the critical concentration $10^{\frac{b}{a}}$, but which embodies no definite theory as to the union.

Variations in K_m for the Same Enzyme.

All preparations of yeast invertase not only hydrolyse sucrose to glucose and fructose, but raffinose to melibiose and fructose, and the two processes are identically affected by pH, except on the acid side of the optimum, where K_m for sucrose at least varies with pH. Nevertheless in ·138 M solutions the ratio of the velocity of hydrolysis of sucrose to that of raffinose varied, in the case of different yeasts, from 5·1 to 12·3, which led to the view that the enzymes concerned were present in different proportions in them. The true explanation was found by Kuhn [1923, 1], who measured the values of the Michaelis constants and limiting velocities for the two sugars. If we represent the v, x relation by $v_s = \frac{V_s x}{K_s + x}$, and $v_r = \frac{V_r x}{K_r + x}$, for the two sugars, we have the following typical results for different yeasts :—

TABLE VII.

v_s/v_r (·138 M).	K_s.	K_r.	$\frac{K_r}{K_s}$.	V_s/V_r.
4·8	·016	0·24	15	2·0
5·0	·016	0·27	16	1·9
8·3	·040	0·66	17	1·9

Hence for all yeast invertases so far investigated, sucrose has about sixteen times the affinity of raffinose, and the compound, when formed, is hydrolysed about twice as fast. This is as near as the errors of observation allow to the ratio of 1·6 found for acid hydrolysis, and suggests that the mechanism may be similar. Now at a concentration of ·14 M the enzyme is always saturated with sucrose, whilst its degree of saturation with raffinose depends on the value of K_r, and thus the ratio of the velocities of hydrolysis is different. Also the inadequacy of invertase to raffinose is determined more by low affinity than by low hydrolysing power. When the same enzyme attacks two substrates the quantitative differences will, in general, depend both on differences in affinity and in velocity of hydrolysis when saturated.

The proportionality of the affinities of sucrose and raffinose may be compared with the proportionality, for the same hæmoglobin, of the affinities for O_2 and CO when both are altered by varying pH or otherwise, found by Douglas, Haldane, and Haldane [1912]. On the other hand, the ratio of these affinities is the same for all yeast saccharases, though different for different hæmoglobins. In this respect, and with regard to the relative values of V, β-glucosidases differ more among themselves than invertases. Thus Willstätter, Kuhn, and Sobotka [1923] found the values of Table VIII.

TABLE VIII.

	Salicin.		β-phenyl-glucoside.		β-methyl-glucoside.	
	K_m	V.	K_m	V.	K_m	V.
Bitter almonds *a* .	·035	8·5	·065	1·06	·60	1
,, ,, *b* .	·035	7·3	·040	·84	·40	1
Apricot . .	·041	9·3	—	—	·65	1
Sweet almonds .	·017	6·8	—	—	1·12	1
Strong acid . .	—	4·3	—	10·0	—	1

The maximal velocities for β-methyl glucoside are arbitrarily taken as unity. It will be seen that the sweet almond emulsin has a greater affinity for salicin and a less for β-methyl glucoside than has that of bitter almonds. It is just possible that the different ratios of V for the different enzyme preparations are due to experimental error. Similar results were found for the hydrolysis of maltose, α-methyl and α-phenyl glucosides by Willstätter, Kuhn, and Sobotka [1924]. Here the ratios of the values of V for α-methyl glucoside, α-phenyl glucoside and maltose were as 1 : 28 : 140, the corresponding values for $\overset{+}{H}$ catalysis being 1 : 25 : 180, that is to say, the same within the limits of experimental error.

In general, purification of invertase does not affect its K_M. In one case [Kuhn, 1923, 1] the affinity was diminished by the addition of boiled invertase solution. This has been variously ascribed to a real alteration in the nature of the enzyme particles, and to the addition of an inhibitory substance.

The Union of the Enzyme with Compounds Related to its Substrate.

It was early found that, even when an enzyme reaction was not reversible, it might be slowed down by the addition of its products. This cannot, in most cases, be due to a back reaction, for the velocity of this is negligible in such cases as that of sucrose hydrolysis. It is sometimes due to pH changes, as in the case of unbuffered urease solution, but it is usually due to different causes. As a matter of fact the classical case of invertase still remains to be fully investigated. The reaction catalysed appears [1] to be—

$$\alpha\ (1,\ 5)\ \text{glucosido-}\beta\ (2,\ 5)\ \text{fructoside} + H_2O \rightarrow \alpha\ (1,\ 5)\ \text{glucose} + \beta\ (2,\ 5)\ \text{fructose.}$$

Subsequently the β (2, 5) fructose changes into an equilibrium mixture of α and β (2, 6) fructose, and part of the α (1, 5) glucose mutarotates more slowly into β (1, 5) glucose. The changes have been followed polarimetrically by Hudson [1909] after rapid enzyme hydrolysis at 0°. Now the old experiments on the influence of reaction products refer entirely to equilibrium mixtures of normal (6-membered ring) glucose and fructose. Later the effects of α- and β-glucose, and of αβ- and β-fructose were separately determined. But the unstable (2, 5) or γ-fructose, which is the immediate product of reaction, has an inhibitory effect no greater than normal fructose according to Nelson and Bodansky [1925].

[1] Cf. Appendix to Chap. VI.

An added compound may affect the velocity in two different ways. It may reduce V, without altering K_m. In this case the velocity is equally reduced at all substrate concentrations. Thus Michaelis and Pechstein [1914, 2] found that 1·0 M glycerol reduced the velocity of inversion of ·05 M and ·20 M sucrose solutions to 40 per cent. and 45 per cent. respectively, while α-methyl glucoside behaved similarly. Presumably such inhibitors, apart from slight effects due to dilution of the water, must unite with the enzyme in such a way as to reduce or abolish its catalytic activity, but are not displaced from it by the substrate. They must therefore be attached to a different part of its surface from the substrate. If the reaction be reversible, and the compound quite devoid of catalytic power, as appears to be the case with that of invertase and α-methyl glucoside, we have

$$\frac{v'}{v} = \frac{1}{1 + \frac{h}{K_h}}$$

where v is the velocity without inhibitor, v' the velocity with it, h the concentration of inhibitor, and K_h the dissociation constant of the combination. This formula is followed provided h does not rise too high, when strong solution effects may also intervene. Fructose, on the other hand, produces a far greater percentage inhibition in weak than in strong sucrose solutions. If we assume, then, that the sucrose and fructose molecules compete for the same spot in the enzyme molecule, and if f be the concentration of fructose, and q that of enzyme-fructose compound, we have the three reactions

$$E + S \rightleftharpoons ES; \quad E + F \rightleftharpoons EF; \quad \text{and } ES + H_2O \rightarrow E + F + G.$$

If K_f be the dissociation constant of EF,

∴ (as on p. 39)

$$K_m p = x(e - p - q), \text{ and } K_f q = f(e - p - q),$$

$$\therefore p = \frac{ex}{x + K_m\left(1 + \frac{f}{K_f}\right)}$$

$$\therefore v' = \frac{Vx}{x + K_m\left(1 + \frac{f}{K_f}\right)} = \frac{v}{1 + \frac{fK_m}{K_f(x + K_m)}},$$

where v is the velocity in the absence of fructose, v' in its presence.

Hence the net effect of a competing substance is to increase K_m, and the amount needed to double it is equal to the dissociation constant K_f. This may also be calculated from the equation

$$K_f = \frac{fK_m}{\left(\frac{V}{v'} - 1\right)x - K_m} = \frac{fK_m}{\left(\frac{v}{v'} - 1\right)\left(x + K_m\right)}.$$

The agreement between values so obtained is satisfactory, considering the errors involved [Kuhn and Münch, 1927]. In Nelson and

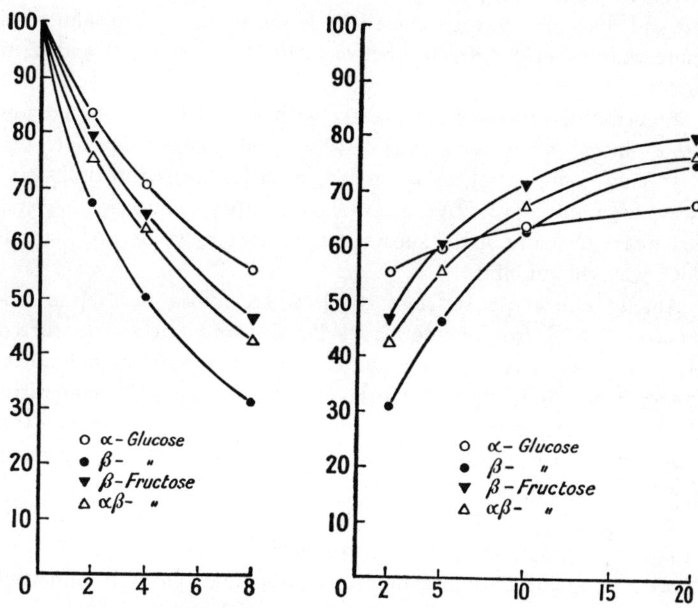

FIG. 21.—Effect of varying glucose and fructose concentrations on initial velocity of hydrolysis of 2 per cent. sucrose by yeast saccharase. Abscissa, hexose concentration per cent. Ordinates, velocities of hydrolysis of sucrose.
[After Nelson and Anderson, 1926.]

FIG. 22.—Effect of varying sucrose concentration on initial rate of its hydrolysis by yeast saccharase in presence of 8 per cent. glucose or fructose. Abscissa, sucrose concentration per cent. Ordinates, velocities of sucrose hydrolysis as percentages of velocity in absence of hexose.
[After Nelson and Anderson, 1926.]

Anderson's [1926] experiments (see Figs. 21 and 22) β-glucose had about half the affinity of sucrose for the enzyme, fructose a somewhat less value. Older data give similar results, though it is not absolutely clear whether all the competing affinities vary proportionally from one enzyme to another. However, Kuhn and Münch's [1927] results make this rather unlikely.

A few substances, e.g. α-glucose in the case of some invertases, and methyl and ethyl alcohols in the case of emulsin, according to Josephson [1925], act in both ways. With sufficient data, it is, however, quite possible to disentangle the two effects.

The literature regarding the combination of inhibitory substances with invertase is voluminous, but it is exceedingly misleading owing to the following facts :—

1. The earlier authors ignore the difference between the action of α and β sugars, first discovered by Kuhn in 1923 ; and almost all ignore the probable difference between the effects of (2, 5) and (2, 6) fructose.

2. Yeast invertases differ greatly with regard to their behaviour with α-glucose, some being inhibited by it, while others are not.

3. There is a general confusion, long after Michaelis and Pechstein's paper, between competitive and non-competitive inhibition. Hence in the case of many of the substances considered, we do not know in which way they inhibit.

Among the most satisfactory data are those of Nelson and Anderson [1926], from whose results Figs. 21 and 22 are constructed. For their enzyme α-glucose inhibited in an almost non-competitive manner, while β-glucose, α-fructose, and αβ-fructose were competitive inhibitors. Since the effect of αβ-fructose is slightly greater than that of β-fructose, α-fructose would be still more active, probably about as active as β-glucose.

The following substances [1] do not inhibit yeast saccharase except perhaps to a small extent in high concentrations :—

Maltose, gentiobiose, cellobiose, lactose, melibiose, tetramethyl-β-methyl glucoside, gluconic acid, amygdalin, lactic acid.

The following inhibit non-competitively :—

α-glucose and trehalose (in some cases), α-methyl glucoside, salicin, glycerol, αβ-mannose (? both components), ethyl alcohol,[2] α-*l*-arabinose.

The following inhibit competitively :—

β-glucose, α-fructose, β-fructose, α-galactose, β-galactose, β-*l*-arabinose.

The following inhibit, according to which law is not certain :—

α-rhamnose, β-rhamnose, α-xylose, β-xylose, β-methyl glucoside, mannitol, hexose-diphosphoric acid.

The cases of other poisons, e.g. metals, halogens, and organic

[1] Euler and Josephson [1924, 1], Kuhn [1923, 2], Kuhn [1924], Kuhn and Münch [1927].

[2] Only as a diluent.

bases, will be dealt with in Chapter VIII. Of the sugars and hexosides the α-form is the more active in the following :—

Fructose, xylose, methyl glucoside ;

and the β-form in the following :—

Galactose, arabinose, mannose, ? lactose, rhamnose.

It is noteworthy that galactose may inhibit more strongly than glucose or fructose.

Nelson and Bodansky [1925] compared the rates of hydrolysis of sucrose (*a*) alone, (*b*) in presence of "nascent" invert sugar, as formed during hydrolysis, and (*c*) in presence of equivalent amounts of mutarotated invert sugar. (*a*) was always greatest, but (*b*) exceeded (*c*) in the early stages of the reaction, (*c*) exceeding (*b*) later on. It follows that α- (2, 5) fructose is probably less inhibitory than αβ- (2, 6) fructose, and may be much less so. The question, however, requires further study.

It is noteworthy that α-methyl glucoside may inhibit in some cases when α-glucose does not. No clear laws can at present be laid down as to the influence of structure on affinity. It is possible that such may emerge when the type of inhibition is determined in each case. We do not know, moreover, whether, for example, α-glucose and α-methyl glucoside unite with the enzyme at the same spot, though this could be determined. The importance of structure is, however, obvious. Josephson [1924] found that the inhibitory effect of fructose increased slowly with the pH, rising by 31 per cent. between pH 2·75 and 6·75. That of glucose was maximal at the optimum, falling to 76 per cent. at pH 2·75, and 60 per cent. at 6·65. Kuhn and Münch [1927] showed that the inhibition by α-glucose and α-galactose was nearly independent of pH, that of β-glucose being maximal at the optimum.

The Affinity of a Synthesizing Enzyme for its two Substrates.

Although the hydrolyses catalysed by enzymes are really bimolecular, the concentration of H_2O cannot be substantially altered without bringing in strong solution effects, and there is so far no experimental ground for supposing that it unites with any enzyme. Where the reverse reaction is measurable, the concentrations of both substrates can be altered. This has been done by Josephson [1925] for prunase, the β-glucosidase component of almond emulsin. Using a large excess of alcohol he found that for glucose $K_m = 0{\cdot}18$ and

0·185 for β-methyl glucoside synthesis in 30 per cent. and 40 per cent. alcohol, and 0·19 for β-ethyl glucoside synthesis in 30 per cent. alcohol at the optimal pH. Owing to the slowness of the reaction the data are of course for equilibrium αβ-glucose, not for pure β-glucose. When the methyl alcohol concentration was varied, matters were complicated by strong solution effects. Half the maximum velocity was reached at a concentration of 3·4 M, but the true value of K must be somewhat higher. Hence the enzyme has a very small affinity (of the competitive type) for methyl alcohol.

Now, in the case of the enzymatic hydrolysis of β-glucosides, glucose exerts a retarding influence, mainly but not wholly due to the β component. The dissociation constants of the enzyme—αβ-glucose combination, deduced from the inhibition of the hydrolyses of salicin, helicin, and arbutin respectively, were 0·16, 0·15, and 0·20, in extremely satisfactory agreement with the result (0·185) found directly. This constitutes a very strong argument for the validity of the theory by which the figures were calculated.

At the optimum pH Josephson found the values of K_m given in Table IX.

TABLE IX.

Alcohol.	β-Glucoside.	K Alcohol.	K Glucoside.	Ratio.
Methanol	β-methyl glucoside	3·7	·71	5·2
Phenol	β-phenyl glucoside	·21	·050	4·2
Saligenin	Salicin . . .	·17	·030	5·7
Salicylaldehyde	Helicin . .	·10	·016	6·2
Quinol	Arbutin. . .	·09	·042	2·1

The figures for the glucosides are obtained by measuring the initial velocities at different concentrations, those for the alcohols by inhibition. It is clear that the two sets of figures are intimately connected. The affinity of the enzyme for quinol is 2 to 3 times that expected from the affinity of its glucoside. This is not due to the quinol having two alcoholic groups, for its monomethyl ether has a K of about ·10. Nevertheless, it is clear that the enzyme has a high affinity for phenols and a much smaller for aliphatic alcohols. The affinity for a glucoside is no doubt also a function of the affinity for its sugar residue, and it would be of very great interest to study quantitatively the behaviour of such sugars as maltose, *d*-epirhamnose, and the methylated glucoses, together with their β-compounds towards this enzyme. Josephson found that maltose (like cane-sugar, lactose, raffinose, and fructose) caused no appreciable inhibition, although galactose, xylose, and arabinose had a small effect. The affinity of the β-maltosides should therefore be low. If K_s, K_a, K_b be the dissociation constants of the compounds of the enzyme with its substrate, and with the two products of reaction, and V_1, V_2 the velocities of hydrolysis and synthesis when the enzyme is saturated, while K is the dissociation constant of the equilibrium reached in the catalysed reaction, it is obvious from a consideration of the velocities in a dilute solution, that $\frac{K_a K_b}{K_s} = \frac{KV_2}{V_1}$ (see p. 82). If, then, V_1 and V_2 were proportional, for different sets of substrates, to the velocity constants in the absence of a specific catalyst (which is often roughly true), then $\frac{K_a K_b}{K_s}$ should be a constant for any given enzyme.

The inhibitory compounds showed both competitive and non-competitive inhibition. It was possible, however, to correct for the latter, and the figures given, which are not very accurate, relate to it only. Salicylate gave only non-competitive inhibition. Incidentally, Josephson's work proves as conclusively as is at present possible the identity of the enzyme hydrolysing the different β-glucosides.

pH had a marked effect on the affinity of this enzyme for its substrate. Thus the dissociation constant of the salicin-enzyme compound rose from ·031 at the optimal pH of 4·4 to ·046 at pH 6·5, and ·044 at pH 2·8. Hence in a dilute solution the velocity falls off more sharply on each side of the optimum than in a strong solution, and it is probably for the same reason that the velocity of hydrolysis of β-methyl glucoside, with which saturation of the enzyme has never been reached, falls off more sharply than that of the aromatic glucosides. It is clear that Michaelis' formula for the relation of velocity and pH can only be expected to hold either if the enzyme is saturated with substrate, or if K_m does not vary with pH.

Acceleration by Products of Hydrolysis.

This phenomenon is generally due to changes of pH, but one case has been observed of a more direct action. Nelson and Cohn [1924] found that in strong sucrose solutions the rate of hydrolysis by honey (bees' salivary) invertase at first increased with time. Nelson and Sottery [1924] found that this was due to the glucose and fructose liberated, mainly the former. Fig. 23 shows the effects on initial rate of hydrolysis of equilibrium mixtures of α, β-glucose and α, β-fructose added to 10 per cent. sucrose at pH 5·76. It will be seen that the maximum acceleration is produced by about 0·6 per cent. glucose and 0·1 per cent. fructose. α-glucose is less efficient, either at acceleration or retardation, than β-glucose, while β-fructose and αβ-fructose have the same effect. In 3 per cent. sucrose the maximum acceleration by glucose occurs on adding between 0·1 and 0·2 per cent., though the acceleration is not so great. The optimum concentration of glucose is therefore roughly proportional to that of sucrose. A similar case occurs outside the field of enzyme chemistry, where Douglas, Haldane, and Haldane [1911] found that, while large amounts of O_2 displace CO from its combination with hæmoglobin, small amounts increase the amount combined. On the other hand, this only occurs with CO concentrations insufficient to saturate the hæmoglobin at all completely, while the enzyme, according to Michaelis' theory, should be very nearly saturated by a 5 per cent. solution of sucrose.

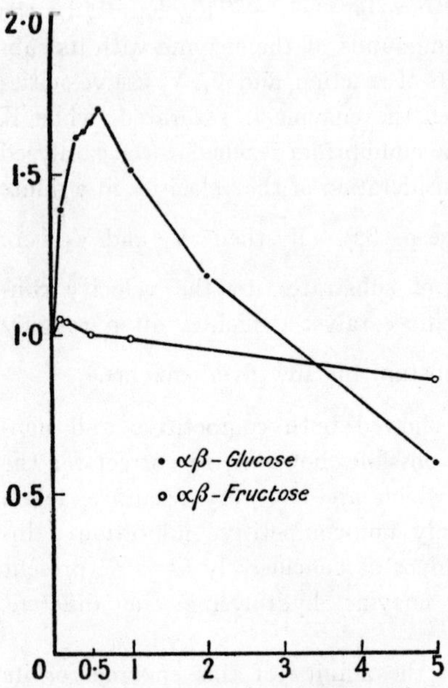

FIG. 23.—Effect of varying glucose and fructose concentration on velocity of sucrose hydrolysis by honey saccharase. Abscissa, hexose concentration per cent. Ordinates, initial rate of hydrolysis of 10 per cent. sucrose.

[After Nelson and Sottery, 1924.]

Lipase.

The substrate concentrations at which certain lipases reach half their maximum velocity of hydrolysis are given in Table V. It will be noticed that these vary very greatly. Moreover, they differ in the case of optical antipodes, showing that a purely physical interpretation is impossible. There are similar differences in the falling off of reaction velocity in high substrate concentrations. Where the enzyme acts on an emulsion, either in a non-aqueous phase or at the phase boundary, the concentration cannot be determined satisfactorily. Murray [1929] investigated the hydrolysis of mixtures by pancreatic lipase. He points out that the affinity of this enzyme is greater for tributyrin than for dibutyrin and monobutyrin [Terroine, 1910] and confirms this for triacetin and monoacetin. The affinity for triacetin also appears to be greater than that for ethyl butyrate.

Murray also investigated inhibition. While fluorides inhibit non-competitively, a number of organic compounds inhibit in a competitive manner. The ratio of their affinities to that of ethyl butyrate is given in Table X.

TABLE X.

Acetophenone	4·35
Phenyl-methyl-carbinol, Benzaldehyde	2·75
Benzophenone	2·5
Phenol, Anisole	1·25
Cyclohexanol	0·60
Aniline	0·45

Other ketones had a moderate effect. Hydrocarbons had very little, and salts of propionic, lactic, and pyruvic acids had no more than sodium chloride. It is also known that glycerol has only a slight inhibitory effect, if any, since glycerol extracts of organs often have a powerful lipase action. It appears probable, from the powerful inhibitory effect of the carbonyl group, that the enzyme unites with the carbonyl group of the substrate. Since acetophenone oxime has no inhibitory effect, the carbonyl group of the inhibitor is clearly responsible. The effect of secondary alcohols is not so clear. Possibly the enzyme unites with an enolic tautomer of its substrate, or the secondary alcohol may unite with the enzyme in a different manner.[1] Willstätter, Kühn, Lind and Memmen [1927] found a latent period of the order of an hour in the hydrolysis of ethyl mandelate by liver lipase. This was traced to the presence of ethyl phenylglyoxylate, $C_6H_5—CO—CO—O—C_2H_5$, as an impurity. Until this substance was hydrolysed, no hydrolysis of the mandelate occurred. This is clearly due to the strong affinity of the enzyme for a carbonyl group in an uncharged molecule, the ionized keto-acid being ineffective, like pyruvic acid. The keto-ester is very slowly hydrolysed, in comparison with ethyl mandelate. The most plausible explanation of this fact is that it is largely united with the enzyme by the wrong carbonyl group, and therefore does not undergo hydrolysis.

[1] Murray and King (1929) have since shown that the affinities of the lævo forms of three secondary alcohols for sheep liver lipase are four to five times as great as those of their dextro isomers.

Pepsin and Trypsin.

Northrop has investigated the action of these proteases very thoroughly. As regards substrate concentration the results for trypsin are entirely different according as the total action, as measured by change in conductivity or production of amino nitrogen, or the actual destruction of casein, is taken as a measure of reaction velocity. In the former case the initial velocity is independent of the substrate concentration over a wide range, in the latter it increases with the concentration of casein [Northrop, 1922, 3]. This result may be due to the fact that all his trypsin solutions contained carboxypolypeptidase and erepsin, or to the fact that trypsin has a lower affinity for casein than for its reaction products. In the case of pepsin [Northrop, 1920, 3], the *v, s* curve agrees very well with the Michaelis-Menten equation, putting $K_m = 4{\cdot}5$ per cent. for egg albumin. As the rate of digestion of casein (as measured by amount of substance precipitated by trichloracetic acid) is unaffected by the addition of 3 per cent. gelatin, which is above the concentration needed to reach a maximum velocity (as measured by conductivity or amino N determinations), he concludes against the formation of a trypsin-protein compound [Northrop, 1922, 4]. The same conclusion is reached because, although a further increase of protein concentration beyond 4 per cent. in the case of casein and 3 per cent. in that of gelatin does not increase the rate of digestion as measured by conductivity (but *not* by a precipitation method), yet the rate in 4 per cent. casein + 3 per cent. gelatin (Fig. 27) is greater than in either alone, though less than their sum [Northrop, 1922, 4]. To the writer this argument would be conclusive were it not that under the conditions of the experiment peptidases as well as trypsin were present, and would act on intermediate products. It is very desirable that the work should be repeated with a purified enzyme.

In the course of digestion both trypsin and pepsin produce inhibitory substances [Northrop, 1920, 2, 1922, 1] which slow down the reaction. The inhibition is non-competitive. Northrop [1922, 2] regards this as evidence against the union of enzyme and protein. To the writer it only seems to prove that the enzyme does not unite with protein and inhibitor at the same point on its surface. The velocities for any given concentrations of enzyme and inhibitor agree very satisfactorily with the hypothesis that there is an equilibrium, Enzyme + inhibitor $\rightleftharpoons$ enzyme-inhibitor compound (Figs. 1, 24). Hence, if an enzyme preparation containing inhibitor be diluted, the compound dissociates, and such preparations possess a greater relative activity in dilute solutions. But, whereas in all other cases investigated (except xanthine-oxidase), the proportion of the inhibitor combined with the enzyme is negligible, so that a given amount of inhibitor reduces the reaction velocity by an amount independent of the enzyme concentration, this is not so here, as is clear from Fig. 24. When the amount of enzyme is increased, the absolute amount bound by the inhibitor is of course increased, but the percentage bound diminishes. Alkali inactivated (but not heat inactivated) pepsin can enter into the equilibrium, inactivated trypsin cannot.

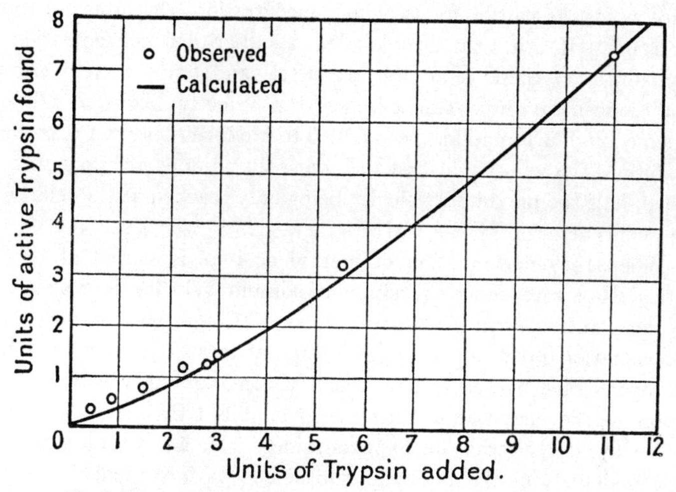

Fig. 24.—The influence of the total amount of trypsin on the inactivation caused by 5 units inhibitor. Increased amounts of trypsin were added to series of tubes each containing 25 c.c. gelatin solution, and 5 units inhibitor. Duplicate series were run at the same time and under the same conditions, but without inhibitor. [Northrop, 1922, 2.]

The inhibitors are dialyzable, but are not amino-acids. The pepsin inhibitor is destroyed by trypsin. The trypsin inhibitor is produced by tryptic, but not by acid or alkali hydrolysis of proteins [Northrop, 1922, 1].

Xanthine-oxidase.

Xanthine-oxidase catalyses the oxidation of xanthine and hypoxanthine, and to a less extent of adenine and acetaldehyde, in presence of a great variety of hydrogen acceptors (see p. 123), including methylene blue. Dixon and Thurlow [1924] found that, while in the case of aldehyde, the maximum velocity is not even approached in 2 per cent. solutions, it is reached in the case of the purine bases in less than ·00003 M concentration. At higher concentration the velocity remains steady up to about ·001 M, the concentration being independent of that of methylene blue but increasing with the strength of the enzyme solution.[1] It then falls off, at first rapidly, and then more slowly (Figs. 17, 18). This inhibition is also exhibited by adenine and uric acid, of which the latter is not oxidized, the former slightly if at all. Dixon and Thurlow call the concentration at which the velocity begins to fall off the critical concentration. In sub-critical concentrations hypoxanthine reduces methylene blue twice as fast as xanthine. At high concentrations the rates are about the same. If uric acid be added to a sub-critical concentration of hypoxanthine, there is at first no effect, and only when a critical joint concentration is reached does the velocity begin to fall off. When enough uric acid has been added to produce roughly a maximum inhibition, the velocity at first rises with the amount of hypoxanthine added, half the maximum velocity being only reached at the relatively high concentration of ·0001 M (Fig. 25).

When the concentration of methylene blue is varied at a sub-critical substrate concentration a maximum velocity is reached at a concentration somewhat under ·00001 M, but with a very high concentration (·002 M) of hypoxanthine at least ten times as much methylene blue is needed to reach the maximum velocity. The action of the enzyme on aldehyde is inhibited by uric acid in the same way as its action on hypoxanthine. This is a strong argument for the identity of the aldehyde oxidase and purine oxidase.

Dixon and Thurlow assumed that the enzyme activates purines at a rate independent of their concentration (at least from ·00003 M upwards), and that as the purine concentration increases it becomes increasingly difficult for the methylene blue to reach the enzyme surface. The following discussion of the experimental facts is quoted (by kind permission) from their paper :—

"1. The hypoxanthine and methylene blue are in all probability adsorbed on the enzyme side by side and there react. At low purine concentrations the enzyme is readily accessible to the methylene blue and therefore the reaction proceeds at a velocity which is determined by the rate at which the hypoxanthine is activated by the enzyme. At high purine concentrations, however, according to assumption two,

[1] From data very kindly shown to me by Dr. Dixon, it appears that these two quantities are at least roughly proportional over a considerable range.

it becomes difficult for the methylene blue to gain access to the enzyme surface, and the rate of methylene blue supply becomes limited. The reaction, therefore, no longer proceeds at its former rate, but with a velocity limited by the rate at which methylene blue can reach the surface of the enzyme.

"2. The initial constant velocity follows from the first assumption. As long as the methylene blue has free access to the enzyme surface the reaction velocity will depend on the rate of hypoxanthine activation.

"3. According to the above assumptions, as the concentration of purine increases, the methylene blue finds it increasingly difficult to reach the enzyme surface, and a stage is finally reached when its rate of supply to the enzyme is equal to the rate at which hypoxanthine is activated. Any further increase in purine concentration will result in a slowing of the rate of methylene blue supply below this value, and the reaction velocity will then be limited by this rate of supply and not by the rate of hypoxanthine activation. Any increase of purine concentration beyond that at which the two rates are equal will cause inhibition.

"4. The fact that the critical concentration depends on the enzyme concentration indicates that much of the purine in solution is associated with the enzyme, because adding more enzyme causes the purine to be distributed over a larger enzyme surface, and hence so great a resistance is not offered to the passage of methylene blue. Otherwise the addition of more enzyme would not decrease the inhibitory effect of the purine.

"5. Since at low purine concentration the reaction velocity depends on the rate of hypoxanthine activation and the methylene blue has free access to the enzyme, the concentration of methylene blue will not affect the reaction velocity.

"6. Since at high purine concentrations the access of methylene blue to the enzyme is limited, the reaction velocity will depend upon the rate of access, and it is to be expected that this will depend on methylene blue concentration.

"7. When the velocity is limited only by the constant rate of substrate activation which we assume to be the same for hypoxanthine and xanthine, hypoxanthine will reduce methylene blue at twice the rate that xanthine does, since it uses two molecules of methylene blue for its conversion to uric acid, whereas xanthine uses but one. Hence the ratio of their rates of reduction of methylene blue is two to one. When, however, inhibition occurs and the velocity comes to depend upon the rate at which methylene blue reaches the enzyme, this rate will be the same for both xanthine and hypoxanthine, and therefore their rates of reduction will be one to one. That the curves show a slight crossing indicates that hypoxanthine is slightly more effective than is xanthine in causing inhibition.

"At the present state of knowledge of enzyme reactions it is extremely difficult to formulate any theory along physico-chemical lines which would explain either the constant rate of hypoxanthine activation by the enzyme or the blocking of the enzyme surface by purines. The temperature coefficient of the reaction, which has been found to have a value of 2·0, indicates that the activation is a true chemical process and one which is not limited by the rate of diffusion of the substrate to the surface of the enzyme. It seems impossible, however, to picture any mechanism by which the rate of activation can remain independent of the substrate concentration." (*Biochem. J.*, **18,** pp. 983, 984.)

While I believe that the above arguments are substantially correct, the following comments may be made. It is not absolutely necessary to assume that the methylene blue ever reaches the enzyme surface, provided it can reach an activated purine base molecule. Langmuir [1916] holds that when CO reacts with O_2 at a platinum surface the CO molecules are oxidized on collision with activated O_2 molecules, but not with the platinum itself. The experiments summarized in Fig. 25 show that uric acid not only prevents methylene blue from reaching its goal, but also hypoxanthine. It is a little difficult to see why the critical concentration should be so sharp. On the hypothesis that at the critical purine concentration the rate of supply of methylene blue is just equal to that at which hypoxanthine is activated, one would expect the critical concentration to be raised by increasing that of methylene blue, which is not the case. It seems possible that at the critical concentration a unimolecular layer of purine base is completed over the enzyme surface, or a certain portion of it, and that beyond this concentration a second layer begins to cover the activated molecules.

Xanthine-oxidase is so far unique among enzymes in that a certain substrate concentration is associated with each concentration of enzyme. Actually 1 milligram of Dixon and Kodama's [1926] most active enzyme solution would be brought to a critical concentration by about 1·9 mg. of hypoxanthine. This is only a possible unimolecular layer if the above preparation is nearly pure, and the enzyme molecule small, flat, or long and thin. As such impurities as casein and fat do not affect the critical value, it would appear likely that the substrate is adsorbed by, or combined with, the enzyme alone.

On any hypothesis it is clear that this enzyme adsorbs, or in some way unites with, more substrate molecules than it can activate. In the case of the hydrolytic enzymes it is quite possible that the same may be true (see p. 84). If so, of course the Michaelis equation only refers to activated molecules.

Citric acid oxidase [Bernheim, 1928, 2] behaves like xanthine-oxidase, but has a larger K_m (about ·0005 M), and a critical concentration about ·01 M.

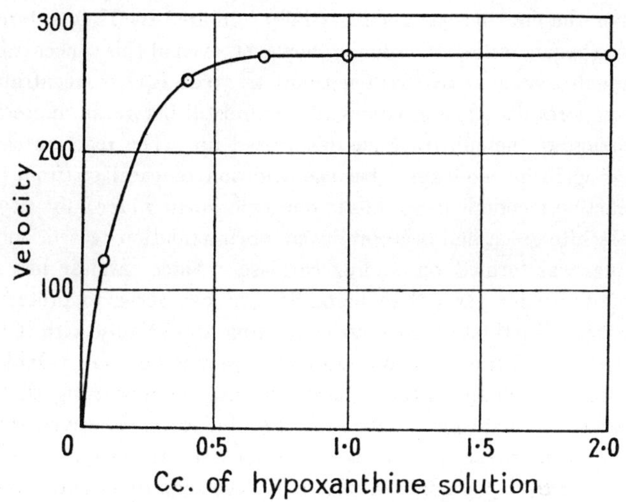

Fig. 25.—Milk xanthine-oxidase. Hypoxanthine concentration curve in presence of uric acid. Each tube contained: 5·0 c.c. enzyme solution (in buffer); 0·5 c.c. methylene blue; 2·0 c.c. uric acid (5 mg. per c.c.); x c.c. hypoxanthine (1 mg. per c.c.); $(2 - x)$ c.c. water. [Dixon and Thurlow, 1924.]

Peroxidase.

In high H_2O_2 concentrations the activity of peroxidase is less than in moderate concentrations. Willstätter and Weber [1926, 1] found that the optimum concentration of H_2O_2 was about 0·25 mg. per 100 c.cm. (·00007 M). The inhibition in higher concentrations has often been attributed to irreversible destruction of the enzyme, but Willstätter and Weber [1926, 2] found that, although this may occur in high H_2O_2 concentrations, there are also two distinct types of reversible inactivation. If the enzyme is mixed with H_2O_2 before adding the reducing substrate (pyrogallol or leuco-malachite green) its activity is reduced to from 90 to 50 per cent. of that observed when both are added simultaneously. This inhibition is independent within wide limits of the time of exposure to H_2O_2. It can be nearly completely reversed by adding larger amounts of pyrogallol. The authors attribute it to the existence of a stable and an unstable type of enzyme peroxide addition product, perhaps with H—O—O—H and $\begin{matrix} H \\ H \end{matrix}\!\!>\!O = O$ respectively.

In relatively low H_2O_2 concentrations (50 mg. per 100 c.cm.) the reaction proceeds at constant velocity for an appreciable time, whether the enzyme has been partially inhibited by H_2O_2 before the reaction began or not. Gradually, however, even at this concentration, and much sooner at five to ten times as great H_2O_2 concentrations, the reaction falls off, and comes to a standstill before an appreciable proportion of the substrate has been used up. The reaction can be started again in two ways. On the addition of catalase (from liver) the reaction recommences; thus in one experiment where only ·015 mg. of malachite green had been produced when oxidation ceased, another ·235 mg. was formed on adding catalase. Since catalase has a far lower affinity for H_2O_2 than peroxidase, it can serve to protect the latter from inactivation without competing appreciably with it when the H_2O_2 concentration as an oxidizing agent is optimal. Oxidation will also start again, after a latent period, on increasing the concentration of reducing substrate. In this case, however, higher substrate concentrations are needed than when the enzyme has been inactivated by H_2O_2 alone, so that it is necessary to assume a second type of inactive enzyme peroxide addition product. As a result, when high H_2O_2 concentrations are used, the velocity of reaction is a function of the concentration of reducing substrate. But it does not follow that the latter combines with the enzyme.

Bacterial Dehydrogenases.

Quastel and Whetham [1925, 1, 2] compared the rates of reduction of methylene blue by different concentrations of various substrates in presence of a suspension of *Bacillus coli*. However high the substrate concentration they did not find the velocity entirely independent of the substrate concentration, though this was nearly the case with the sugars. With most of the substrates examined, it was found that the relation between velocity and substrate concentration was

$$v = k + k' \log s = k' \log \left(\frac{s}{c}\right).$$

Clearly $v = 0$ when $s = c$, i.e. no action occurs below a certain critical substrate concentration. Table XI. shows these critical concentrations for certain substrates, together with concentrations giving the half-maximal velocity, as in Table V. The latter are

TABLE XI.

Substrate.	Critical Concentration.	"Half-maximal" Concentration.
Formic acid	·005	·0003
Acetic acid [1]	·2	·1
Propionic acid [1]	< ·2	—
Butyric acid [1]	·3	·17
Glycollic acid	·1	·15
Lactic acid	about ·0007	·007
Glyceric acid	—	·1
Alanine	—	·07
Succinic acid	about ·0005	·01
Glutamic acid	—	·002
Xylose	003	·01
Galactose	·0005	·005
Fructose	<·00007	·0001
Glucose	<·00007	<·0001

obviously rough, owing to the indefinite nature of the maximum. All concentrations are molar. When no critical concentration is given, it appears to be either zero, or at least far lower than the "half-maximal" concentration.

It is, of course, out of the question that all the acetic acid in a ·2 M solution should have been adsorbed elsewhere than on the active parts of the bacterial surface. Hence the phenomenon is really an expression, in the main at any rate, of a property of the enzymes themselves. As it is also found with malt amylase, it may be a very general property hitherto masked by the use of too high concentrations in most cases.

It is clear that several different catalysts are concerned in these oxidation-reduction processes. That catalysing reduction by lactic acid has been brought into solution by Stephenson [1928] and Quastel and Wooldridge [1928]. The solution catalyses the oxidation of lactic and α-hydroxybutyric acids, and related compounds (see Chap. V.). *B. coli* after treatment with toluene, which inactivates certain of its catalysts, behaves very similarly to this particular enzyme. It is competitively inactivated by oxalic acid.

The following acids strongly inhibit lactic acid reduction :—

[1] The possible presence of formic acid as an impurity may affect these figures.

α-hydroxybutyric, glyceric, pyruvic, glycollic, glyoxylic, mandelic, oxalic, hydroxymalonic, mesotartaric, parabanic.

The following, among others, do not inhibit :—

Acetic acid, glycol, malonic acid, succinic acid, hydantoin.

Other substances, such as glycine, glycerol, and *d-l*-tartaric acid are slightly inhibitory.

The strong inhibitors all possess the structure

—CO—COH or —CHOH—COH,
/\ /\

the H on the right being ionizable, and may be regarded as analogous either to pyruvic or lactic acid.

Similarly the enzyme dehydrogenating succinic acid is strongly inhibited by the following acids :—

Malonic, glutaric, pyrotartaric, mesotartaric, tricarballylic, phenyl-propionic.

Malonic acid, which has a very high affinity, inhibits competitively.

When, however, the untreated organism is studied the properties of the lactic acid dehydrogenase at any rate are markedly different. The lactic acid dehydrogenating activity of the toluene treated organism is slightly, but perhaps significantly, above that of the normal. Now, whereas the velocity of methylene blue reduction by $\frac{M}{140}$ lactic acid is reduced to 9 per cent. of its original value by $\frac{M}{112}$ oxalic acid in the case of toluene treated *B. coli*, it is only reduced to 54 per cent. in the normal organism. Quastel and Wooldridge [1927, 2] showed that this result is not due to altered permeability. They conclude that the toluene alters the relative affinities of the enzyme for lactic and oxalic acids. It is also conceivable that the effect of toluene is double. It inactivates an enzyme or group of enzymes which have a relatively low affinity for oxalic acid (and, from other considerations, are probably fairly non-specific) and increases the activity of one particular enzyme, which has a high affinity for lactic acid. It would seem that some activation by the toluene must in any case be assumed, but Dr. Quastel has suggested to me that the slight accelerating effect of toluene may be due to its removing an adsorbed film of O_2 on the bacteria. The possibility must, however, be kept in mind that in this case a treatment which has a small effect on the V of an enzyme may greatly alter its K_m.

These surface catalysts are more fully dealt with by Stephenson in *Bacterial Metabolism* (1930).

Results with some other Enzymes.

Wohl and Glimm [1910] found the following inhibitions with malt amylase :—

5 per cent.	Glucose,	49 per cent.
10 ,,	,,	100 ,,
10 ,,	Mannose,	15 ,,
10 ,,	Fructose,	0 ,,
10 ,,	Maltose,	67 ,,
15 ,,	,,	100 ,,
15 ,,	Sucrose,	0 ,,
15 ,,	Dextrin,	75 ,,

These figures apply to maltose production. The effects on starch hydrolysis, as measured by the disappearance of the iodine reaction, are far slighter. Ambard [1923] showed that the polyamyloses behave in the same way. Henri and Philoche [1904] found that fructose and glucose inhibit *Aspergillus* maltase, the latter more intensely.

In general, substances related to the substrate produce inhibition when present in concentrations of the same order of magnitude as that of the substrate, i.e. their affinities are of the same order of magnitude. This is strikingly shown by Martland and Robison [1927] in the case of bone phosphatase, which has as great an affinity for its substrate as any hydrolytic enzyme so far investigated. Here the velocity of hydrolysis of 0·1 M glycerol phosphate was halved by about ·026 M phosphate, while even ·1 M glycerol had no effect. The same phosphate concentration had about the same effect on ·03 M glycerol phosphate, so that the inhibition appears to be mainly non-competitive, as in the case of α-glucose and invertase. Mutch [1912] found that the action of histozym was inhibited by sodium benzoate, but not by glycine.

Euler and Josephson [1926] found that glycine inhibits the hydrolysis of glycyl-glycine by erepsin only on the alkaline side of neutrality. They concluded that iso-electric glycine has no effect, only the anion combining with erepsin. The inhibition is competitive. Alanine behaves in the same manner. Glycine anhydride (dioxo-piperazine) which is not hydrolysed, and a peptone which is very slowly hydrolysed, also inhibit, while urea, aceturic acid, and benzoyl-glycyl glycine, which are not hydrolysed, do not inhibit. Nucleosidase, acting on adenosin, is inhibited to the extent of 46 per cent. by an equi-molecular concentration of nucleic acid, and to the extent of

22 per cent. by a somewhat stronger solution of uric acid; but not at all by arabinose [Euler and Brunius, 1927].

Protection by the Substrate.

In presence of their substrates and related bodies, enzymes are generally protected to a certain extent against inactivation by heat or chemical agents. Thus Bayliss and Starling [1904] showed that trypsin is protected from heat by boiled egg albumin or peptone, Wohl and Glimm [1910] found that malt amylase was protected by starch, some but not all samples of dextrin, maltose, glucose, fructose, and sucrose. But some of these sugars, such as sucrose, do not inhibit starch hydrolysis. Hence the protection does not imply a union of the enzyme with sugars of the same type as that responsible for inhibition. Similarly, Ter Meulen [1905] found that at 51° C. glucose and lactose both protected emulsin, but only the former inhibited it. We must therefore distinguish at least three types of union between an enzyme and its substrate or similar bodies:—

(*a*) Causing catalysis or competitive inhibition.

(*b*) Causing non-competitive inhibition.

(*c*) Causing protection from heat or chemical agents, but no inhibition or catalysis.

Examples of protection from chemical influence are given in Chapter VIII.

CHAPTER IV.

THE INFLUENCE OF TEMPERATURE AND RADIATION ON ENZYME ACTION.

TEMPERATURE has two effects, reversible and irreversible. To study the latter the enzyme is raised to a certain temperature for a known time, and then brought back to a lower temperature at which it is stable, and compared with a specimen which has not been heated. It is always desirable that this should be done before any temperature coefficients of the catalysed reaction are measured. Where a reaction is substantially reversible, the equilibrium will, of course, vary with temperature. This, however, though it complicates the results in a few cases, throws no light on the nature of enzymic catalysis.

The Temperature Coefficients of Enzyme Actions.

Table XII. gives the temperature coefficient Q_{10}, i.e. the ratio of the velocities at $(T + 10)^\circ$ and T° for a number of enzymes. The data for yeast saccharase, which has been more thoroughly studied, are considered later. The table is far from complete, as many of the data in the literature are clearly unreliable. Thus Auld [1908] obtained values for emulsin hydrolysis of amygdalin varying quite irregularly between 1·30 and 3·61 at seven different temperature intervals. Moreover, different authors obtain different values for the same reaction. Thus, Terroine [1910] found Q_{10} of 2·03 and 1·96 for the hydrolysis of ethyl butyrate by pancreatic lipase between 0° and 15° and 15° and 26°, in conflict with Kastle and Loevenhart [1900]. It will be noticed that most of the values are below 2·0, i.e. lower than those for the majority of chemical reactions. On the other hand, very high figures were found for trypsin (other authors have found values in the neighbourhood of 3·0). Moreover, the coefficient of hydrolytic, but not oxidative enzymes, almost invariably falls off with temperature. This may be due to several causes. Thus at 39° trypsin is already undergoing irreversible inactivation under certain conditions. Kastle and Loevenhart's lipase systems were possibly macroheterogeneous. In such a case the rate of reaction is partly regulated by diffusion, and as the temperature rises this process will assume an increasing importance, since, having a lower temperature coefficient than most chemical reactions, it will tend to become the slowest process of the series whose end result is being measured. Nevertheless, the case of yeast saccharase makes it clear that the falling off with temperature is a real phenomenon.

TABLE XII.

TEMPERATURE COEFFICIENTS OF SOME ENZYME ACTIONS.

Enzyme.	Substrate.	Temp.	Q_{10}.	Reference.
Pancreatic lipase	Ethyl butyrate	0°-10°	1·50	Kastle and Loevenhart (1900).
		10°-20°	1·34	
		20°-30°	1·26	
Liver lipase	,, ,,	0°-10°	1·72	
		10°-20°	1·36	
		20°-30°	1·10	
Yeast maltase	Maltose	10°-20°	1·90	Lintner and Kröber (1895).
		20°-30°	1·44	
		30°-40°	1·28	
Malt amylase	Starch	20°-30°	1·96	Lüers and Wasmund (1921).
		30°-40°	1·65	
		40°-50°	1·43	
Yeast erepsin	Glycyl-tyrosine	25°-35°	2·25	Abderhalden, Caemmerer, and Pincussen (1909).
		35°-45°	1·56	
Pancreatic erepsin	,, ,,	25°-35°	1·67	
		35°-45°	1·60	
Pepsin	Edestin	0°-10°	2·6	Ege (1925).
		10°-20°	2·0	
		20°-30°	1·8	
		30°-40°	1·6	
		40°-45°	1·4	
Trypsin	Casein	21°-31°	5·3	Bayliss (1904).
		31°-39°	3·3	,, ,,
Urease	Urea	30°-40°	3·0	Van Slyke and Cullen (1914).
Fat catalase	H_2O_2	0°-20°	1·4	Nordefeldt (1920).
Xanthine-oxidase	Hypoxanthine	—	2·0	Dixon and Thurlow (1924).
Root peroxidase	Leuco-malachite green	5°-15°,	2·0	Willstätter and Weber (1926, 1).
		15°-25°	2·0	
B. coli succinoxidase	Succinate	30°-40°	2·1	Quastel and Whetham (1924).
		40°-50°	2·1	
		50°-60°	2·1	

In all cases the temperature coefficient of an enzymatic reaction has been found to be less than that of the same reaction when "uncatalysed" or catalysed by $\overset{+}{H}$ ions Thus at 20° the energies of activation or critical increment in calories per gram-molecule, calculated from the equation

$$E = \frac{R \log_e \frac{k_2}{k_1}}{\frac{1}{T_1} - \frac{1}{T_2}}$$

for sucrose hydrolysis at 20° by acid and saccharase respectively are 25,500 and 9400. The corresponding figures for ester hydrolysis are 17,400, and 4100 (Kastle) or 11,600 (Terroine). This must be the case if we regard the catalyst as so straining the substrate molecule as to lower its heat of activation. The phenomenon is, of course, common to all catalysts.

Yeast saccharase has been most thoroughly studied by Nelson and Bloomfield [1924], who have combined all the data in the literature, including those of Euler and Laurin [1920], who arrive at similar conclusions. At the optimum pH the apparent energy of activation

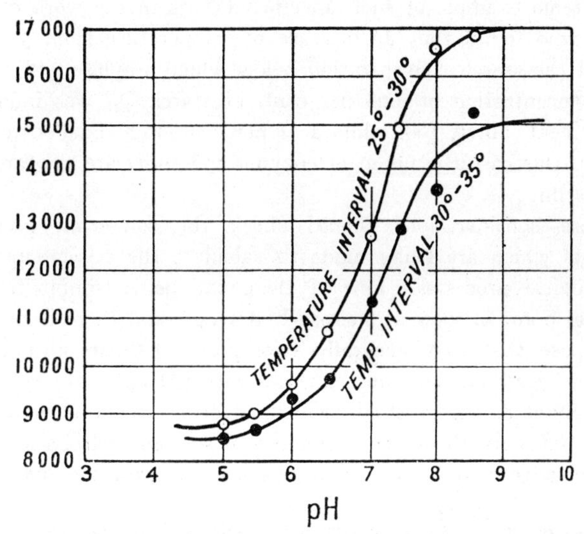

FIG. 26.—Apparent energy (in calories per gram-molecule) of activation of saccharase-saccharose compound as a function of pH. [Nelson and Bloomfield, 1924.]

varies from 11,400 at 5·6° to 5800 at 48·8°. The results agree with the equation $E = 11{,}400 - 100t$, t being the centigrade temperature. For a short distance on each side of the optimum pH the values are fairly steady, but they rise violently on the alkaline side (Fig. 26), and fall off on the acid side (though here the phenomenon is partly but not wholly due to irreversible inactivation). The change on the alkaline side is due to the variation of the pH-velocity relation (see p. 20) with temperature. Thus at pH 8, when the temperature is raised from 30° C. to 35° C., two things happen. The velocity at the optimum pH is increased by 28 per cent., and, owing to a shifting of the pH-activity curve towards the alkaline side, the velocity at pH 8 becomes 9·1 per cent. of that at the optimum, instead of 7·3 per cent. Hence the velocity is increased by 58 per cent., giving a spurious value for E of 16,500 calories. In the case of saccharase K_m has a very small temperature coefficient, but this may not be universally the case. If it is not, we should expect to obtain the true energy of activation only in high substrate concentrations. Sub-maximal substrate concentrations should give a spuriously low or high temperature coefficient, the increase in the velocity of transformation of the enzyme-substrate compound being partly counteracted by the decreased formation of the compound or augmented by its increased formation. An extreme example of such an effect occurs in the work of Eadie [1926], who found that an increase of temperature from 5° to 15° lowered the starch concentration below which amylase will not act. At a concentration of ·068 per cent. of starch Q_{10} was infinite, at ·07 per cent. about 100, while at 1 per cent. it had fallen to about 2·5. In this case the union of enzyme and substrate appears to be endothermic.

These considerations should check the somewhat premature attempts which are being made to calculate the critical increment of biological processes. Few of them can be as simple from the chemical point of view as sucrose hydrolysis, and there is no reason to suppose that they generally take place at their optimum pH. Thus the yeast cell contains enzymes with pH optima varying from 4·6 to about 8, and most of them must act at a non-optimal pH.

On Michaelis' theory the velocities of sucrose hydrolysis over the alkaline range in sufficient concentrations of sucrose are given by the equation $v = \frac{V[\overset{+}{H}]}{[\overset{+}{H}] + K_a}$, where K_a is the acid dissociation constant of saccharase and of its sucrose compound. Nelson and Bloomfield [1924] found $K_a = 10^{-7}$ at 25°, and $·675 \times 10^{-7}$ at 35°, so that the Q_{10} of the reaction is 1·48, and the molar heat of dissociation of the hydrogen ion 7200 calories. The dissociation constant near 10^{-2} appears to vary in the same way. Hence neither can be due to a carboxyl group. On the other hand, both may well be due to the ionization of aromatic or heterocyclic amino groups. It may be remarked that in most cases where temperature coefficients have been measured (though apparently not in the above measurements on saccharase) the pH was varying with temperature owing to varying dissociation of the buffer.

Inactivation by Heat.

Most enzymes are very rapidly destroyed in aqueous solutions below the boiling-point, although for example, according to Miyake and Ito [1924] *Aspergillus oryzae* amylase solution, suitably prepared, retains some activity even after heating at 100° for two hours, and some other vegetable enzymes are relatively heat-stable. Dried preparations are usually far more stable; for example dried rennin is only slowly destroyed at 158° C. The course of inactivation is sometimes that of a unimolecular reaction, as in the case of dry rennin according to Arrhenius, and dissolved emulsin, rennin, pepsin, and trypsin, according to Tammann [1895], and Arrhenius [1907]. This is not always the case. Thus Euler and Laurin [1919] found that in the case of yeast saccharase the unimolecular reaction "constant" fell off during the reaction.

The most striking characteristic of heat inactivation is its enormous temperature coefficient, which may be larger than that of any other reaction except the inactivation of lysins. Thus Zilva [1914] found $Q_{10} = 3040$ for milk peroxidase at 70°. Some values of the critical thermal increment are given in Table XIII. The values found are comparable with those of the reactions shown below,

TABLE XIII.

	A.	T_c°.	
Lipase (dry) . .	26,000	151	Nicloux (1904).
Emulsin (dry) .	26,300	101	Tammann (1895).
,, (wet) .	45,000	54	,, ,,
Malt amylase . .	42,500	57	Luers and Wasmund (1922).
Trypsin . . .	62,000	65	Arrhenius (1907).
Pepsin . . .	75,000	65	,, ,,
Rennin . . .	90,000	46	,, ,,
Saccharase (pH 4·0).	101,000	59	Euler and Laurin (1919).
Peroxidase . .	189,000	69	Zilva (1914).
Hæmoglobin (denaturation) .	77,500	63	Lewis (1926, 1).
Egg albumin (denaturation) .	130,000	76	,, ,,
Vibriolysin . .	129,000	46·5	Famulener and Madsen (1908).
Goat hæmolysin .	198,500	50·5	,, ,, ,,

and have been regarded as a support for the protein-like nature of enzymes. However, purified enzymes are generally much more thermolabile than crude preparations, and the increase in lability often coincides with the disappearance of protein reactions (see Chap. IX.). The critical temperature and temperature coefficient are certainly sometimes, and possibly always, those of protective proteins or other colloids associated with the enzyme, and not properties of the enzyme itself. It is intelligible that a reaction involving a large molecule should need a large quantum of energy. The critical inactivation temperature, given in the third column, is that at which the enzyme is half destroyed in an hour. Owing to the large temperature coefficients, this temperature is pretty sharply defined.

Both A and T_c are affected by pH. The variation of the latter can be seen from Fig. 10. In this case A is also a maximum near the optimum, falling off to only 53,000 at pH 5·7. The different values in Table XIII. may, in part, be due to such causes. Euler and Laurin found that A had the same values for a top and bottom yeast, the latter, which normally lives at 25°, having T_c about 2° C. higher. The difference was not due to heat-stable protective substances, since when inactivated, it did not protect the more thermolabile saccharase.

Miyake and Ito [1926] found that *Aspergillus* amylase which had been partially inactivated by heating, recovered its activity to a small extent on standing in the cold. Thus, after an hour at 85° the activity was reduced to 46 per cent., but on standing for four days it had risen to 55 per cent. of the initial activity. This was probably due to enzyme carried down by protein coagulum going into solution again. Most enzymes are not rapidly destroyed at the temperature of liquid air.

The Optimum Temperature.

As the temperature is raised the catalysed reaction proceeds more rapidly, but in addition the velocity falls off more rapidly owing to inactivation of the enzyme. At a certain temperature these two effects will balance. However, the optimum is not definite, and will be higher the smaller is the amount of work to be done. Thus a given quantity of amylase may just destroy the iodine-blueing properties of a starch solution at 55° in ten minutes, this being its optimum temperature. But if the experiment is repeated on twenty times the quantity of starch, the enzyme will be largely inactivated before the end of 200 minutes, and the process will take a much longer time, or never be completed at all. If the temperature is lowered and the experiment repeated, the reaction will proceed somewhat more slowly. But inactivation, which has a very high temperature coefficient, will proceed a great deal more slowly, and hence the hydrolysis of the starch will be completed earlier. Where the reaction measured is completed in much less than an hour, the optimum temperature slightly exceeds the critical temperature of inactivation (in presence of substrate). Where the time taken is measured in hours it will lie somewhat below the critical temperature. The optimum temperature, then, is a practically convenient expression with no absolute significance, and where no other data are available it gives a rough idea of the heat stability of the enzyme, but it is not a definite characteristic even of a given enzyme preparation, much less of the enzyme itself.

Alleged Activation by Heat.

Euler and Blix [1919] found an irreversible increase in catalase activity in yeast cell suspensions by four to nine times on heating them for an hour to 55°-60° C., and attributed it to the destruction of an inhibitory substance. Euler and Borgenstam [1920] found the same phenomenon with blood catalase. But Nosaka [1928, 3] has shown that this is due to hæmolysis. The corpuscular surface catalase differs from the intracorpuscular enzyme in its properties, being, for example, more thermolabile, but when the latter is liberated, the net effect is an increased rate of H_2O_2 destruction.

The Action of Ultra-Violet Radiation on Enzymes.

It has long been known that ultra-violet radiation generally, and visible radiation sometimes, influence enzyme action. Usually the effect is an irreversible destruction, occasionally an activation. The rate of destruction, as was to be expected, is practically independent of temperature. But it is affected by pH and other chemical conditions, in particular presence of the substrate, optical sensitizers, and oxygen.

Whereas enzymes are generally stablest to heat inactivation near their optimal pH, the reverse is often the case with regard to ultraviolet radiation. Thus Seligsohn [1926] found that blood catalase unirradiated was more active at pH 6·8 than at 6·2 or 7·3, while after irradiation it was less so. Similar results were obtained by Pincussen [1924, 1] with malt diastase and Uehara [1928] with pepsin. Other ions may have a great effect. Thus Kumanomidoh [1928] found that while salivary amylase was equally active in various chloride solutions in the absence of radiation, it was protected from radiation by $CaCl_2$, though only slightly by other chlorides. Enzymes are generally more affected in dilute than in concentrated solution and become less stable on purification. Pincussen [1924, 3] found this in the case of amylase and trypsin. Thus a crude pancreatic amylase preparation showed from 0 to 3 per cent. loss of activity after eighteen minutes' irradiation. A purer preparation under the same conditions lost 27 per cent. activity. Diluted to half the former concentration it lost 39 per cent., and again diluted to one quarter, 50 per cent. The substrate may afford considerable protection. The nature of the inactivation is often far from clear. For example, Pincussen [1926] found that malt amylase almost inactivated by ultra-violet radiation could be brought back to near its original activity by the addition of unirradiated enzyme.

In general these experiments have been made in presence of air, however Jamada and Jodlbauer [1908] and Tappeiner and Jodlbauer [1908] found that even in the absence of O_2, peroxidase and catalase were inactivated by ultra-violet radiation. On the other hand, Pincussen [1924, 2] found that the destructive effect of ultra-violet radiation on malt amylase was greatly enhanced by the presence of KI, which liberates I_2 when radiated. But pancreatic, salivary, and *Aspergillus* amylases are on the whole protected by iodides, as by other salts. Dreyer and Hanssen [1907] found that the course of inactivation of a number of enzymes was that of a unimolecular reaction; the rate being proportional to the amount of radiation absorbed. But in other cases significant deviations from such a law have been observed.

The Action of Visible Radiation on Enzymes.

Reynolds Green [1897] investigated the action of light in various parts of the visible spectrum on salivary amylase. He found that red and (to a less extent) blue light produced an activation, which in the former case exceeded 50 per cent. It may be remarked that his preparation showed an increased activity of this order on standing. Green, and more markedly, violet light, produced partial inactivation, less than that due to ultra-violet. Activation by visible radiation is often dependent on the presence of O_2. This was found by Tappeiner and Jodlbauer [1925] for saccharase and amylase, by Jamada and Jodlbauer [1908] and Zeller and Jodlbauer [1908] for peroxidase and catalase. This phenomenon was thoroughly studied by Dixon and Bernheim [1928]. They found that light acting either on the enzyme solution or the methylene blue subsequently added to it, produced at first an acceleration, and then a retardation of the catalysed reduction of methylene blue by hypoxanthine in presence of xanthine-oxidase, so long as O_2 was present. This was traced to the production of H_2O_2. It was found that the activity was about doubled by 10^{-6} M H_2O_2, the activation requiring about two hours. With concentrations of 10^{-3} M and over the enzyme was partly destroyed. The optimum concentration of H_2O_2 is roughly proportional to the strength of the enzyme solution. Other similar results are discussed by these authors.

Jamada and Jodlbauer [1908], Zeller and Jodlbauer [1908], and Tappeiner [1908] described the action of fluorescent substances on zymase, peroxidase, saccharase, and catalase. In general these substances, even when without action in the dark, destroy the enzymes in the light. It is not clear whether this is due to induced oxidation, but it is improbable, since the destruction is rather specific. Thus eosin and rose Bengal sensitized all the enzymes studied, while methylene blue and sodium dichloranthracene-sulphonate sensitized catalase, but not peroxidase, which was actually protected in some cases.

The Action of β-, γ-, and X-Rays on Enzymes.

Hussey and Thompson [1922-1926] studied the effects on trypsin, pepsin, and yeast saccharase of radium emanation kept in their neighbourhood. Evidence was obtained that the inactivation found in each case was mainly due to the "β-rays," i.e. high velocity electrons. A smaller effect (per unit of energy absorbed) was found with γ-rays. In the case of X-rays no effect was obtained in some cases, a distinct effect in others. The effect of X- and γ-rays may be due to the electrons ejected by them. In the case of radioactive sources they found, if Q was the amount of enzyme present, $\frac{1}{Q}\frac{dQ}{dW}$ = constant, where W is the average activity of the radioactive source, multiplied by the time over which it acts. In other words, a given amount of energy, or a given number of β particles, inactivate a definite fraction of the enzyme present.

CHAPTER V.

THE COURSE OF ENZYMATIC REACTIONS, AND ITS MATHEMATICAL THEORY.

A VAST amount of work has been done on the kinetics of reactions catalysed by enzymes. Much of it was done under conditions so complicated as to throw little light on the mechanisms involved. Thus whereas numerous authors had found that lipase action falls off with time, Rona and Lasnitzki [1924] and Rona and Ammon [1927], using respectively a manometric method measuring infinitesimal changes, and a titration method in which pH was kept constant, found the velocity to remain constant for long periods.

If an enzyme obeys the Michaelis equation, then at constant pH three cases may arise :—

(*a*) The substrate concentrations remain so high until very near the end of the experiment that the enzyme is fully saturated. This is the case with many enzymes of high affinity, including some lipases and oxidases.

(*b*) The substrate concentrations are so low that the amount of enzyme combined is proportional to the substrate concentration. This is true in many experiments with enzymes of low affinity.

(*c*) The substrate concentration may vary so that neither of these conditions is fulfilled.

If a be the initial substrate concentration, y the amount transformed in time t, and other symbols as in Chapter III., we have :—

(*a*) If the products of enzymatic action do not inhibit, $y = Vt$; if they inhibit, then the velocity falls off with time. Various equations have been constructed to meet such cases. If the affinities are low in comparison with that of the substrate, as is usually the case, the falling off is less than that characteristic of a unimolecular reaction.

(*b*) In this case, if the products do not inhibit $\frac{dy}{dt} = \frac{(a-y)V}{K_m}$,

$$\therefore t = \frac{K_m}{V} \log_e\left(\frac{a}{a-y}\right),$$

i.e. the reaction is unimolecular. If the products inhibit, the unimolecular constant falls off.

(c) In the absence of inhibition $\frac{dy}{dt} = \frac{V(a-y)}{a-y+K_m}$,

$$\therefore Vt = \int_a^y \frac{(a-y+K_m)dy}{a-y}$$

$$= y + K_m \log_e\left(\frac{a}{a-y}\right)$$

$$= y\left(1 + \frac{K_m}{a}\right) + y^2\frac{K_m}{2a^2} + \frac{y^3K_m}{3a^3} + \quad . \quad . \quad .$$

This equation, first formulated by Henri [1903] gives a good approximation to the course of many hydrolyses. If a is large compared with K_m the logarithmic term is at first negligible, and the reaction proceeds uniformly, slowing down when y becomes an appreciable fraction of $\frac{a^2}{K_m}$.

The Course of Saccharose Hydrolysis.

A variety of equations have been produced to fit the course of such reactions as saccharose hydrolysis. All of them neglect the mutarotation which the products undergo, and the fact that in many cases there is non-competitive as well as competitive inhibition. When mutarotation is taken into account, a differential equation is reached which cannot be integrated in finite terms. If we neglect on the one hand mutarotation, and on the other, the non-competitive inhibition by α-glucose, we may make the following calculation (Michaelis and Menten).

The following reactions are considered, E representing enzyme, S saccharose, F fructose, G glucose, the molecular concentrations of each being written below. a is the initial saccharose concentration.

$$\begin{array}{llcll} 1. & S & + & E & \rightleftharpoons SE \rightarrow E + F + G. \\ & (a-y) & & (e-p-q-r) & \quad p \\ 2. & F & + & E & \rightleftharpoons FE \\ & y & & (e-p-q-r) & \quad q \\ 3. & G & + & E & \rightleftharpoons GE \\ & y & & (e-p-q-r) & \quad r \end{array}$$

It is supposed that in the first reaction the reactions $S + E \rightleftharpoons SE$ and $SE \rightleftharpoons S + E$ are very rapid compared with $SE \rightleftharpoons E + F + G$.

Then if K_m, K_f, K_g are the dissociation constants of SE, FE, and GE,

$$(a - y)(e - p - q - r) = K_m p$$
$$y(e - p - q - r) = K_f q$$
$$y(e - p - q - r) = K_g r.$$

Hence, eliminating q and r

$$p = \frac{e(a - y)}{a - y + K_m\left(1 + \frac{y}{K_f} + \frac{y}{K_g}\right)},$$

$$\therefore \frac{dy}{dt} = \frac{V(a - y)}{a - y + K_m\left(1 + \frac{y}{K_f} + \frac{y}{K_g}\right)},$$

where V is the initial velocity in a strong solution,

$$\therefore Vt = \left(1 - \frac{K_m}{K_f} - \frac{K_m}{K_g}\right) y + K_m\left(1 + \frac{a}{K_f} + \frac{a}{K_g}\right) \log_e \frac{a}{a - y}.$$

Clearly the same equation may be extended to a number of inhibitory products. It leads to a curve of the same nature as does Henri's equation, and gives a very fair fit to such curves as those of Fig. 13. Actually Nelson and Hitchcock [1921] found that the course of hydrolysis was best represented by the equation

$$kt = \log \frac{1}{1 - y} + \cdot 2642y - \cdot 0886y^2 - \cdot 1034y^3,$$

where y is the proportion of sucrose hydrolysed in a 10 per cent. solution. The last two terms represent deviations from the equations of Henri or Michaelis. They are probably, as Nelson and Bodansky [1925] point out, mainly due to mutarotation. However, non-competitive inhibition by α-glucose will account for the third term.

Enzyme Destruction.

An enzyme may be destroyed or inactivated during the course of a reaction. This happens in all cases near the optimum temperature, and also at relatively low temperatures where H_2O_2 is a substrate (as with catalase and peroxidase) or a reaction product (as with xanthine oxidase).

The simplest case is when the enzyme is being destroyed by heat, the proportion destroyed per minute being constant, and the rate of catalysis by the remainder being independent of the falling substrate concentration. This is often nearly true with enzymes of high affinity. If x be the amount of substrate at time t, V the original velocity,

$$\therefore -\frac{dx}{dt} = Ve^{-kt},$$

where k is a positive constant giving the rate of destruction of the enzyme. Then if a be the original substrate concentration,

$$x = a + \frac{V}{k}(e^{-kt} - 1).$$

Putting t infinite, $x = a - \frac{V}{k}$. Hence if $V \geqslant ak$ the reaction is completed, if $V < ak$ the reaction stops after a quantity $\frac{V}{k}$ of substrate has been converted. Hence the total amount of substrate changed varies as the amount of enzyme added, although the latter is a catalyst. Such cases have occurred, and led to considerable confusion. Thus Bach and Chodat [1904] found that the amount of pyrogallol oxidized by H_2O_2 under certain circumstances was proportional to the amount of peroxidase (an enzyme of very high affinity). Oppenheimer [1910] regarded this as contradicting the view that peroxidase is an enzyme. Similarly in the case of the Michaelis equation we have

$$a - x + K_m \log_e \frac{a}{x} = \frac{V}{k}(1 - e^{-kt}).$$

Hence the reaction always stops before it is complete, but if k is not too large it may be nearly completed.

For small values of k, $x_\infty = ae^{\frac{-V}{kK_m}}$ approximately.

Generally, however, either the substrate protects the enzyme, or destroys it. In the former case let E and a be the initial concentrations of enzyme and substrate, y and x their concentrations at time t, and p that of the enzyme-substrate compound. Let K_m be the Michaelis constant, k the velocity constant of the inactivation of the free enzyme, l that of the breakdown of enzyme-substrate compound which is measured, so that the initial velocity V at high substrate concentrations $= El$.

$$\therefore K_m p = x(y - p), \quad \frac{dx}{dt} = -lp, \quad \frac{dy}{dt} = -k(y - p),$$

$$\therefore \frac{dy}{dt} = -\frac{kK_m y}{K_m + x}$$

$$\frac{dx}{dt} = -\frac{lxy}{K_m + x} = \frac{lx}{kK_m} \cdot \frac{dy}{dt},$$

$$\therefore \log_e x = \frac{l}{kK_m} y + \text{constant}.$$

But when $x = a, \quad y = \mathrm{E},$

$$\therefore x = ae^{\frac{l(y - \mathrm{E})}{k\mathrm{K}_m}}$$

$$\therefore y = \mathrm{E} - \frac{k\mathrm{K}_m}{l} \log_e \frac{a}{x}.$$

Hence y vanishes and the reaction stops when

$$\log_e \frac{a}{x_\infty} = \frac{l\mathrm{E}}{k\mathrm{K}_m} = \frac{\mathrm{V}}{k\mathrm{K}_m}.$$

Putting

$$\frac{\mathrm{V}}{k\mathrm{K}_m} = q, \quad x_\infty = ae^{-q}.$$

Hence the reaction is never completed, and the proportion transformed when equilibrium is practically reached is $1 - e^{-q}$ or $1 - e^{\frac{l\mathrm{E}}{k\mathrm{K}_m}}$. It therefore increases with the amount of enzyme, being proportional to it when it is small, but is independent of the substrate concentration, a conclusion which has not, I believe, been verified experimentally. During the reaction

$$\frac{dx}{dt} = \frac{-lx}{x + \mathrm{K}_m}\left(\mathrm{E} + \frac{k\mathrm{K}_m}{l} \log_e \frac{x}{a}\right),$$

$$\therefore lt = \int_x^a \frac{(x + \mathrm{K}_m)dx}{x\left(\mathrm{E} + \frac{k\mathrm{K}_m}{l} \log_e \frac{x}{a}\right)}, \text{ or if } z = \frac{xe^q}{a},$$

$$k\mathrm{K}_m t = \int_{\frac{xe^q}{a}}^{e^q} \frac{(\mathrm{K}_m + aze^{-q})dz}{z \log_e z},$$

$$\therefore kt = \frac{a}{\mathrm{K}_m e^q}\left[\mathrm{E}i(q) - \mathrm{E}i\left(q \log_e \frac{x}{a}\right)\right] - \log_e\left[1 - \frac{\log_e \frac{a}{x}}{q}\right]$$

where

$$\mathrm{E}i(z) \equiv \int_{-\infty}^{z} \frac{e^u du}{u},$$

a function which is tabulated, and

$$q = \frac{\mathrm{V}}{k\mathrm{K}_m}.$$

In order to obtain a theoretical value for the optimum temperature for a given amount x of catalysis it would be necessary to differentiate this expression with regard to the temperature T, assuming V and k to be exponential functions of the temperature. This leads to a highly complicated formula from which, if the numerical data were known, the optimum temperature could be calculated in any given case.

Examples of reactions during which the enzyme is being destroyed are frequent in the literature, e.g. Ter Meulen [1905]. The velocity falls off more rapidly than in the case of a unimolecular reaction.

The Case of Catalase.

Here the enzyme is stabler in the absence than in the presence of substrate. Without H_2O_2 inactivation is inappreciable below 40° C. [Morgulis and Beber, 1928]. In presence of H_2O_2 inactivation is fairly rapid even at 10° C., the most satisfactory data on its catalytic activity being obtained at 0° C. Yamazaki [1921] and Nosaka [1928] concluded that in the case of the catalase of hæmolyzed blood, within certain limits, the rate of destruction both of H_2O_2 and of catalase were proportional to the product of the enzyme and substrate concentrations. If a be the initial H_2O_2 concentration, x its value at time t, b the initial catalase concentration, y its value at time t, we have Yamazaki's equations,

$$\dot{x} = -kxy, \; \dot{y} = -k'xy,$$
$$\therefore \quad k'x - ky = ak' - bk.$$

If this quantity is positive, then at the end of the reaction H_2O_2 is left over, and the amount at any given time is given by

$$\frac{x}{a} = \frac{ak' - bk}{ak' - e^{(bk-ak')t}bk},$$

the final amount being $x = a\left(1 - \frac{bk}{ak'}\right)$. Hence the total amount of H_2O_2 destroyed is $\frac{bk}{k'}$, i.e. is proportional to the amount of catalase. This fact led many authors to doubt whether catalase was really a catalyst.

If $bk > ak'$ the H_2O_2 is all destroyed, and

$$\frac{x}{a} = \frac{bk - ak}{bke^{(bk-ak')t} - ak}.$$

Nosaka [1928 1, 2] found that k (the rate of H_2O_2 destruction) has a maximum at pH 7·0 while k' is little affected by pH. The temperture coefficient of enzyme destruction is much larger than that of H_2O_2 destruction. The rather inadequate data concerning peroxidase can probably be interpreted along similar lines.

Reversible Reactions.

Consider the reactions

$$\underset{(e-p)}{\mathrm{E}} + \underset{x}{\mathrm{S}} \underset{k_2}{\overset{k_1}{\rightleftharpoons}} \underset{p}{\mathrm{ES}} \underset{k_4}{\overset{k_3}{\rightleftharpoons}} \underset{(e-p)}{\mathrm{E}} + \underset{y}{\mathrm{P}}$$

where the total amount of enzyme is e, of substrates $x + y = a$, and the velocity constants of the various reactions are written above or below the arrows.

Then (as on p. 40) we have

$$\frac{dp}{dt} = k_1(e-p)x + k_4(e-p)y - k_2p - k_3p = 0,$$

$$\therefore p = \frac{(k_1x + k_4y)e}{k_2 + k_3 + k_1x + k_4y},$$

the observed reaction velocity,

$$v = k_3p - k_4(e-p)y$$

$$= \frac{e(k_1k_3x - k_2k_4y)}{k_2 + k_3 + k_1x + k_4y}.$$

Hence equilibrium is reached when

$$\frac{x}{y} = \frac{k_2k_4}{k_1k_3},$$

i.e. the above quantity is equal to the equilibrium constant of the reaction, which depends only on the free energies of S and P, and not on the catalyst.

Putting

$$y = 0, \quad v = \frac{k_3ex}{x + \dfrac{k_2 + k_3}{k_1}}.$$

Hence for the reaction

$$\mathrm{S} \rightarrow \mathrm{P}, \quad \mathrm{V} = k_3e, \quad \mathrm{K} = \frac{k_2 + k_3}{k_1}$$

where V is the maximum velocity, K the Michaelis constant. Similarly for the reaction

$$\mathrm{P} \rightarrow \mathrm{S}, \quad \mathrm{V}' = k_2e, \quad \mathrm{K}' = \frac{k_2 + k_3}{k_4}.$$

Now this is not in accordance with the facts. For example, if S is β-methyl-*d*-glucoside, and P glucose, the rate of the back reaction V′ is roughly proportional to the concentration of methyl alcohol in the solution. Clearly this cannot affect the rate of breakdown of the glucoside-enzyme compound into enzyme and glucoside, i.e. k_2. We must therefore develop a completer theory.

Consider the reactions

$$\underset{x}{S} + \underset{(e-p-q)}{E} \underset{k_2}{\overset{k_1}{\rightleftharpoons}} \underset{p}{SE} \underset{k_4}{\overset{k_3}{\rightleftharpoons}} \underset{q}{PE} \underset{k_6}{\overset{k_5}{\rightleftharpoons}} \underset{y}{P} + \underset{(e-p-q)}{E},$$

where the six velocity constants are k_1, k_2, etc., and the concentrations of the various reactants as above. The above equation may be applied to such a case as the reversible hydrolysis of β-methyl glucoside, where S represents glucoside, P glucose, provided the concentrations of water and methyl alcohol are little altered during the reaction. In this case k_3 should be roughly proportional to the concentration of H_2O, k_4 to that of CH_4O.

We have then

$$\frac{dp}{dt} = k_1x(e - p - q) + k_4q - (k_2 + k_3)p = 0,$$

$$\frac{dq}{dt} = k_6y(e - p - q) + k_3p - (k_4 + k_5)q = 0.$$

Solving these equations for p and q,

$$\frac{p}{e} = \frac{k_1(k_4 + k_5)x + k_4k_6y}{k_2k_4 + k_2k_5 + k_3k_5 + k_1x(k_3 + k_4 + k_5) + k_6y(k_2 + k_3 + k_4)}$$

$$\frac{q}{e} = \frac{k_1k_3x + k_6(k_2 + k_3)y}{k_2k_4 + k_2k_5 + k_3k_5 + k_1x(k_3 + k_4 + k_5) + k_6y(k_2 + k_3 + k_4)}$$

$$\therefore v = k_3p - k_4q,$$

$$= \frac{(k_1k_3k_5x - k_2k_4k_6y)e}{k_2k_4 + k_2k_5 + k_3k_5 + k_1x(k_3 + k_4 + k_5) + k_6y(k_2 + k_3 + k_4)}$$

$$= \frac{VK'x - V'Ky}{KK' + K'x + Ky}$$

where

$$V = \frac{k_3k_5e}{k_3 + k_4 + k_5}, \quad V' = \frac{k_2k_4e}{k_2 + k_3 + k_4},$$

$$K = \frac{k_2k_4 + k_2k_5 + k_3k_5}{k_1(k_3 + k_4 + k_5)}, \quad K' = \frac{k_2k_4 + k_2k_5 + k_3k_5}{k_6(k_2 + k_3 + k_4)}.$$

If $y = 0, \quad v = \dfrac{Vx}{K + x},$

hence V is the maximum velocity of the reaction $S \rightarrow P$ in absence of P, K its Michaelis constant, V′ and K′ the same constants for the reverse reaction. We cannot calculate even the ratios of the six k's from these four quantities. However, it is in some cases an observed fact that the velocity v of the reaction varies with influences which can only be supposed to affect k_3 and k_4. Thus Josephson [1925, 2] found that the rate of synthesis of β-methyl glucoside was proportional to the concentration of methyl alcohol, provided this did not exceed 20 per cent. In other words, V′ is proportional to k_4. This can only be the case if k_4 is small compared with k_2, and similarly it is probable that k_3 is small compared with k_5. Probably k_3 and k_4 are smaller than the other constants. If we can take this to be the case we have approximately

$$V = k_3 e, \quad V' = k_4 e, \quad K = \frac{k_2}{k_1}, \quad K' = \frac{k_5}{k_6}.$$

Hence the Michaelis-Menten assumption is applicable, except that the limiting velocity is given, not by the rate of breakdown of the enzyme-substrate compound, but by the rate of its transformation into a compound of the enzyme and another substrate. In the case of sucrose hydrolysis k_4 is infinitesimal, and k_3 is the velocity constant of some such reaction as

$$\text{Saccharase-saccharose} + H_2O \rightarrow \text{Saccharase-(fructose} + \text{glucose)}.$$

The latter complex breaks down rapidly, unless it is reformed; and since $k_5 > k_3$, the amount of it present is negligible unless fructose and glucose are present in the solution in appreciable amounts. It must not, of course, be assumed that k_3 and k_4 are always relatively small. It is difficult to imagine that this is the case with peroxidase.

Perhaps, however, the most important result of the above investigation is that the equation

$$v = \frac{VK'x - V'Ky}{KK' + K'x + Ky},$$

represents the course of the reaction no matter what are the relative values of the velocity constants. When equilibrium is reached

$$\frac{x}{y} = \frac{V'K}{VK'} = \frac{k_2 k_4 k_6}{k_1 k_3 k_5}.$$

So this quantity is equal to the equilibrium constant, which depends on the free energies of S and P, and is independent of the catalyst. This relation is borne out by the facts. Euler and Josephson [1924, 2] calculated the ratio of the affinities of β-glucosidase for glucose and β-methyl glucoside as 4·0 from the velocity constants and Bourquelot's [1914] equilibrium data. The Michaelis constants calculated from their then data gave a ratio 3·3, but Josephson's [1925] data (see p. 49) give K_m for glucose 0·1775 (mean of six different methods), for β-methyl glucoside 0·71. The ratio is therefore exactly 4·0 as expected.

If a is the total amount of substrate S + P, we may write $y = a - x$, and if b is the amount of S at equilibrium, putting $z = x - b$, i.e. the departure from the equilibrium value, we have

$$b = \frac{V'Ka}{V'K + VK''}$$

$$v = \frac{(VK' + V'K)z}{KK'\left[1 + \frac{(V + V')a}{VK' + V'K}\right] + (K - K')z} = -\frac{dz}{dt}.$$

Hence Henri's equation describes the change of z with time. z may be positive or negative, so if z is plotted against t the curve consists of two parts approaching the same asymptote. If K = K', both are logarithmic curves. Otherwise one may have an approximately straight portion, but if so the other cannot. Examples of such pairs of curves obtained in dilute solutions are given by Woolf [1929]. Most of the reversible reactions so far followed have occurred in such strong solutions that their kinetics are relatively complicated.

Bi-molecular Reactions.

The full discussion of these reactions is very complicated. We consider only the case of an irreversible reaction where the enzyme-substrate compound is transformed at a velocity negligible compared with those involved in its formation and breakdown. We have then four equilibria,

$E + A \rightleftharpoons EA$, $E + B \rightleftharpoons EB$, $EA + B \rightleftharpoons EAB$ and $EB + A \rightleftharpoons EAB$.

Let K_1, K_2, K_3, K_4 be their dissociation constants, so that $K_1K_3 = K_2K_4$.

Let k be the velocity constant of the reaction $EAB \rightarrow E + C$.

Let e be the total amount of enzyme, p the amount of EAB.

$$\therefore p = \frac{exy}{K_1K_3 + K_3x + K_4y + xy},$$

$$\therefore v = \frac{kexy}{K_1K_3 + K_3x + K_4y + xy}.$$

Hence if y is constant and x varied,

$$V = \frac{key}{K_3 + y}, \quad K_m = \frac{K_1K_3 + K_4y}{K_3 + y},$$

i.e. if y is small compared with K_3,

$$V = \frac{key}{K_3}, \quad K_m = K_1,$$

and if y is large compared with K_3,

$$V = ke, \qquad K_m = K_4.$$

Hence the Michaelis constant of one of the two substrates may vary with the concentration of the other.

If the fact that the enzyme has united with one substrate does not influence its affinity for the other, then

$$K_4 = K_1, \quad K_3 = K_2, \quad \text{and} \quad v = \frac{kexy}{(x + K_1)(x + K_2)},$$

i.e. the Michaelis constants do not vary in this way. Josephson [1925, 2] has shown that this is roughly true for prunase. Where the enzyme-substrate combination depends on molecular charge, as with the proteases, it is rather unlikely that K_1 and K_4 should be equal. Otherwise the expectation is not unreasonable as a first approximation.

Attempts have been made to calculate the Michaelis constant of a molecule from those of its products of hydrolysis. They have not so far been very successful, perhaps because the importance of H_2O as a reactant has not been sufficiently stressed.

Inhibition by Excessive Substrate Concentrations.

This phenomenon is fairly common, a good example being furnished by the lipases, cf. Fig. 16. The following theory will fit the observations in this case, though not in that of xanthine oxidase.

Consider the reactions

$$\underset{(e-p-q)}{E} + \underset{x}{S} \rightleftharpoons \underset{p}{ES}$$

$$\underset{p}{ES} + \underset{x}{S} \rightleftharpoons \underset{q}{ES_2}$$

$$\underset{p}{ES} \rightarrow E + P.$$

The molecular concentration of each reactant species is written below it. Let K_1, K_2 be the Michaelis constants of the first two reactions, k the velocity constant of the third, supposed to be relatively slow.

$$\therefore (e - p - q)x = K_1 p, \quad px = K_2 q,$$

$$\therefore p = \frac{ex}{K_1 + x + \dfrac{x^2}{K_2}},$$

$$\therefore v = \frac{kex}{K_1 + x + \dfrac{x^2}{K_2}}.$$

The reaction proceeds at a maximum velocity

$$\frac{ke\sqrt{K_2}}{2\sqrt{K_1} + \sqrt{K_2}}$$

when $x = \sqrt{K_1 K_2}$. This is a fraction

$$\frac{1}{1 + 2\sqrt{\frac{K_1}{K_2}}}$$

of the maximum possible if EA_2 were not formed.

Since

$$\frac{v}{ke} = \frac{1}{1 + \frac{K_1}{x} + \frac{x}{K_2}},$$

it is clear that when velocity is plotted against the logarithm of substrate concentration, we should obtain the same familiar bell-shaped curve as when the dissociation residue of an ampholyte is plotted against pH. No work has been done in concentrations high enough to verify the symmetry of the curve, but it is at least nearly symmetrical for sheep's liver lipase, for which in the case of ethyl-*l*-mandelate, $K_1 = \cdot 00166$, $K_2 = \cdot 166$ approximately. In this case the nature of the compound ES_2 may be readily imagined. Whereas in EA the enzyme is united to the ester at two spots (see Chap. X.), one on the alcohol, the other on the acid residue, in EA_2 each molecule is only once united, and hence the strain producing hydrolysis does not arise.

In these cases Bamann and Schmeller [1929] have shown that the velocity increases, sometimes by as much as 27 per cent., during the course of the hydrolysis, in spite of possible back reactions. The equation for the amount x left from an initial concentration a after time t is

$$Vt = a - x + \frac{a^2 - x^2}{2K_2} + K_1 \log_e \frac{a}{x}.$$

When $a - x$ is plotted against t the curve is at first concave upwards, as in an autocatalytic reaction. In relatively high initial concentrations the velocity goes on increasing till the reaction is almost complete.

Competing Substrates and Asymmetrical Hydrolysis.

This case has assumed importance with the investigation by Willstätter, Kuhn and Bamann [1928], and Weber and Ammon

[1929] of the hydrolysis of *d*, *l*, and *dl*-esters by lipases. The theory here given is an extension of theirs. Consider the hydrolysis of a mixture of two substrates S_1 and S_2. Let the initial concentrations be a_1, a_2, the maximal velocities for each substrate alone V_1, V_2, the Michaelis constants K_1, K_2. Let x_1, x_2 be the concentrations of substrate at any moment, p_1, p_2 the concentrations of enzyme combined with each, e the total concentration of enzyme. Consider the case of S_1. We have the reactions

$$E + S_1 \rightleftharpoons ES_1 \rightarrow E + S_1',$$

where S_1' stands for the product of reaction. Then the concentration of free enzyme is $e - p_1 - p_2$, that of ES_1 is p_1, that of S_1 being x_1.

Hence (as on p. 39)

$$x_1(e - p_1 - p_2) = K_1 p_1.$$

Similarly,

$$x_2(e - p_1 - p_2) = K_2 p_2.$$

$$\therefore p_1 = \frac{K_2 x_1 e}{K_2 x_1 + K_1 x_2 + K_1 K_2},$$

$$\therefore \frac{\dfrac{1}{x_1}\dfrac{dx_1}{dt}}{\dfrac{1}{x_2}\dfrac{dx_2}{dt}} = \frac{V_1 K_2}{V_2 K_1},$$

$$\therefore \log \frac{a_1}{x_1} = \frac{V_1 K_2}{V_2 K_1} \log \frac{a_2}{x_2}.$$

Or if

$$a_1 = a_2, \quad \frac{\log x_1}{\log x_2} = \frac{V_1 K_2}{V_2 K_1}.$$

Weber and Ammon show that in the case of Willstätter, Kuhn, and Bamann's hydrolysis of *dl*-ethyl mandelate with pig's liver lipase, where $x_1 + x_2$ is given by the degree of total hydrolysis, $x_1 - x_2$ by the polarimetric reading, the values of $\frac{\log x_1}{\log x_2}$ only vary between 2·9 and 3·1, and that this value agrees fairly well with that of 2·0 calculated from the V and K values, which, however, have a considerable uncertainty.

$$\text{Log} \frac{x_2}{a_2} = \frac{V_2 K_1}{V_1 K_2} \log \frac{x_1}{a_1},$$

$$\therefore x_2 = a_2 \left(\frac{x_1}{a_1}\right)^k, \text{ where } k = \frac{V_2 K_1}{V_1 K_2},$$

$$\therefore -\frac{dx_1}{dt} = \frac{V_1 K_2 x_1}{K_2 x_1 + K_1 a_2 \left(\dfrac{x_1}{a_1}\right)^k + K_1 K_2},$$

$$\therefore V_1 t = \int_{x_1}^{a_1}\left[1 + \frac{K_1}{x} + \frac{K_1 a_2 x^{k-1}}{K_2 a_1{}^k}\right]dx,$$

$$\therefore t = \frac{a_1 - x_1}{V_1} + \frac{a_2}{V_2}\left[1 - \left(\frac{x_1}{a_1}\right)^k\right] + \frac{K_1}{V_1}\log\frac{a_1}{x_1}.$$

This is Henri's equation with an extra term which is clearly small if k is small, and large if k is large. Hence if V_1K_2 and V_2K_1 are very different, the more rapidly destroyed component of the mixture will disappear almost according to Henri's equation. The more slowly destroyed will, under certain circumstances, be destroyed slowly at first, and then reach a maximum rate of destruction. Let A_1 be more slowly destroyed, so that $k > 1$. Then the maximum rate of transformation of A_1 occurs when

$$\frac{x_1}{a_1} = \left(\frac{K_2}{ka_2 - a_2}\right)^{\frac{1}{k}}.$$

So the velocity only increases with time if $a_2 > \frac{K_2}{k - 1}$. In a racemic mixture $a_1 = a_2$, and the optical rotation of the products of hydrolysis is proportional to $x_1 - x_2$, or

$$x_1 - a\left(\frac{x_1}{a}\right)^k.$$

This is a maximum when

$$1 - k\left(\frac{x_1}{a}\right)^{k-1} = 0,$$

$$\therefore x_1 = ak^{\frac{1}{1-k}}, \quad x_2 = ak^{\frac{k}{1-k}}.$$

The maximum rotation is, therefore, proportional to

$$k^{\frac{1}{1-k}} - k^{\frac{k}{1-k}}, \text{ or } k^{\frac{1}{1-k}}\left(1 - \frac{1}{k}\right).$$

When $k = 1$, this is of course zero. As k increases or decreases it gradually increases towards unity, reaching a value $\frac{1}{2}$ when $k = 4{\cdot}5$ or $\cdot\dot{2}$, i.e. in order to obtain during hydrolysis a rotation equal to half that which would be reached were the enzyme absolutely specific, k must be as large as 4·5, or as small as $\frac{2}{9}$. In the case quoted above $k = 3$ approximately, and the maximum rotation should have been over 38·5 per cent. of that theoretically possible with an absolutely specific enzyme. The maximum found was 35·4 per cent.

The proportion of the substrate un-hydrolysed at the time of maximum rotation is

$$\frac{1}{2}\left(k^{\frac{1}{1-k}} + k^{\frac{k}{1-k}}\right).$$

Now when $k = 1$, this $= e^{-1}$ or ·368, and it only increases to ·401 when $k = 5$ or ·2, and to ·5 when k is 0 or ∞. Hence the maximum rotation should always occur at a hydrolysis of between 63·2 per cent. and 50 per cent. of the racemic mixture. Rona and Ammon [1927] in relatively rough experiments, found maximum rotations at percentage hydrolyses varying between 33 and 55 per cent., but Willstätter, Kuhn, and Bamann in very careful experiments found a maximum rotation at 50·5 per cent. hydrolysis, and Bamann [1929] at 52 per cent. They extracted the mandelic acid at different stages of hydrolysis, and took polarimetric readings. The specific rotation of the mandelic acid was plotted against the degree of rotation. The curve so obtained was independent of the pH and enzyme concentration. It should also be independent of the initial substrate concentration, but this was not varied. Its theoretical equation is $1 - x + xy = (1 - x - xy)^k$ where x is the fraction hydrolysed, y the rotatory power of the mandelic acid divided by that of the pure acid. The agreement with prediction was not exact, the rotations observed being too low at both ends of the curve. The rotation of the mandelic acid should be maximal at first, being $\frac{k-1}{k+1}$ of that of d-mandelic acid, and then falling.

Actually it rose from 60·1° to 70·0°, corresponding to $k = 2·6$. This initial rise can perhaps be explained by the partial failure of the Michaelis equation actually observed at high substrate concentrations. The fall at high degrees of hydrolysis is not at present explicable. Bamann [1929] found no effect of altering the initial substrate concentration in the case of pig's liver lipase, but a large effect with human and rabbit preparations. The acid liberated by them at high substrate concentrations was dextro-rotatory, at low concentrations lævo-rotatory.

This cannot be explained in terms of any theory so far put forward. It is clear that our conceptions of enzyme-substrate affinity are still in a primitive stage. The theory given above should hold good, as regards the relation between hydrolysis and optical activity, though not as regards kinetics, even if we allow for inhibition by high substrate concentrations as on page 84.

Consecutive Reactions.

Consider a chain of irreversible catalysed reactions, A → B → C → D, etc., each reaction alone being subject to the Michaelis-Menten conditions. Such examples are frequent in nature. They include the hydrolysis of proteins and triglycerides in dilute solutions, and probably alcoholic fermentation by zymin in presence of excess of phosphate. We will confine ourselves at first to the reactions A → B → C, where the velocity of the first reaction alone is given by $v_1 = \frac{V_1 x}{x + K_1}$, of the second by $v_2 = \frac{V_2 y}{y + K_2}$; x, y, and z being the concentrations of A, B, and C, respectively. Let a be the initial concentration of A. The differential equations for the system are clearly

$$\frac{dx}{dt} = \frac{-V_1 x}{x + K_1}, \quad \frac{dy}{dt} = \frac{V_1 x}{x + K_1} - \frac{V_2 y}{y + K_2},$$

provided A, B, and C are not competing for the same enzyme.

These are not in general integrable except numerically, however they can be integrated provided K_1 and K_2 are small or large compared with a. Four cases arise :—

1. K_1 and K_2 are large. In this case the individual reactions are nearly quasi-unimolecular, and the case is the familiar one of two consecutive reactions, or of the decay of radium emanation.

The two velocity constants are $\frac{V_1}{K_1}$ and $\frac{V_2}{K_2}$.

2. K_1 and K_2 are small. Clearly the rate of destruction of A and production of B is steady till near the end of the reaction, and equal to V_1. Two cases arise :—

(*a*) $V_1 > V_2$. In this case y increases steadily with time at a rate $V_1 - V_2$ until A is nearly exhausted. It reaches a maximum value just less than $\left(1 - \frac{V_2}{V_1}\right) a$, and then falls at a rate V_2, until near the end of the reaction. Its graph consists of two straight lines; z increases at a constant rate V_2 till near the end of the reaction. There is, of course, a small latent period in the production of C.

(*b*) $V_2 > V_1$. The velocity of the first reaction is now the limiting factor. After a short initial lag the value of y reaches the small value $\frac{V_1 K_2}{V_2 - V_1}$, and remains there till almost the end of the reaction.

The rate of production of C is equal to V_1 throughout. This probably represents roughly the state of affairs in many metabolic processes, where the intermediary metabolites are rapidly destroyed, but the rate of the total reaction is fairly steady.

3. K_1 is small, K_2 is large. In this case A diminishes at a constant rate, so that

$$\frac{dx}{dt} = -V_1, \quad \frac{dy}{dt} = V_1 - \frac{V_2 y}{y + K_2}.$$

Two cases arise :—

(*a*) $V_1 > V_2$. $\therefore (V_1 - V_2)t = y - \frac{V_2 K_2}{V_1 - V_2} \log_e \left[1 + \frac{(V_1 - V_2)y}{V_1 K_2}\right]$,

y increases fairly steadily with time, reaching a maximum less than $\left(1 - \frac{V_2}{V_1}\right) a$ when A is nearly exhausted; z increases slowly at first, then more rapidly, and its rate then slows down again. Its graph is nowhere linear.

(*b*) $V_2 > V_1$. Again the velocity of the first reaction is the limiting factor.

$$\therefore (V_2 - V_1)t = y + \frac{V_2 K_2}{V_2 - V_1} \log_e \left[1 - \frac{(V_2 - V_1)y}{V_1 K_2}\right],$$

y comes very near to the value $\frac{V_1 K_2}{V_2 - V_1}$ (unless this is nearly as large as a), and remains there till A is almost exhausted. This value is not necessarily very small. Until this time C is produced at a constant rate V_1, after the lag period is over. On plotting z against t a graph is obtained with a linear central portion, but appreciably bent at both ends.

4. K_1 is large, K_2 small.

$$\therefore \frac{dx}{dt} = \frac{-V_1 x}{K_1}, \quad \therefore x = ae^{\frac{-V_1 t}{K_1}},$$

$$\frac{dy}{dt} = \frac{-dx}{dt} - V_2 \quad \therefore y = a - ae^{\frac{-V_1 t}{K_1}} - V_2 t,$$

x decreases according to Henri's equation, y increases, reaching a maximum value of

$$a - \frac{V_2 K_1}{V_1}\left[1 + \log_e \frac{V_1 a}{V_2 K_1}\right],$$

provided that V_2 is not too large. It soon falls from this maximum again; however, the rate of increase of z is linear and equal to V_2 as long as y is several times as large as K_1. If V_2 is very large y is always less than K_2, and the transformation of A into C almost instantaneous.

In a series of more than two reactions matters are more complicated. When, however, all the reactants have a high affinity for the catalysts concerned, the velocity of the whole is equal to the slowest of the V constants. If we consider these in order V_1, V_2, V_3, etc., it is clear that intermediate products cannot accumulate after the slowest reaction has been accomplished; before this they do so whenever a reaction proceeds more slowly than any of the preceding reactions.

Schutz's Law.

Schutz [1885] found that the amount of peptone produced in a given time by pepsin from egg albumin varied as the square root of the amount of pepsin. This was confirmed by J. Schutz [1900] and others, provided that not more than half the albumin was digested. The reason for a falling off of peptic activity with pepsin concentration was shown by Northrop (see Chap. III.) to be the combination of pepsin with an inhibitory substance. Arrhenius [1907] extended this rule to the tryptic digestion of gelatin and casein, and also to the course of the reaction. He found that over a considerable range the amount of protein digested could be represented by the formula $K\sqrt{et}$ where e is the enzyme concentration, t the time. This formula gives for the velocity of the reaction $v = K\sqrt{\frac{e}{2t}}$. It is at once clear that this formula cannot hold when t is large or small. For the amount digested becomes infinite when t is infinite, the velocity infinite when t is zero. Northrop [1924, 3] investigated the matter thoroughly and found, as was to be expected, that Schutz's law, as modified by Arrhenius, gave too high values when t is small, too low when t is large. Nevertheless, it gives a good approximation to fact over a fairly wide range. But it must be regarded rather as a convenient rule than as a law throwing light on the nature of enzyme action.

The equation $y = k\sqrt{t}$ has been successfully applied to lipase action in unbuffered solutions by Herzog [1913] and Kuhn [1925, 2], who attributes the slowing down to enzyme destruction, the equation $v \propto \sqrt{e}$ to lipase action by Goldschmidt [1925].

General Considerations.

The question often arises as to what deductions can be made from the observed kinetics of an enzyme reaction. If the solutions are unbuffered very little can be deduced unless the substrate and reaction products are non-electrolytes. If the velocity is constant during the utilization of an appreciable fraction of the substrate, we may conclude that the enzyme is all combined with substrate, i.e. that K_m is considerably less than the substrate concentration of the experiment. Moreover, the reaction products cannot unite with appreciable quantities of enzyme under the circumstances of the experiment.

When the velocity falls off any of the following may be occurring :—

1. Inadequate saturation of the enzyme as the substrate concentration falls.

2. Reversible union of part of the enzyme with reaction products.

3. Reversible union of part of the enzyme with substrate to form an inactive compound.

4. Irreversible destruction of the enzyme.

5. Changes of pH.

6. Changes in the state of aggregation of the substrate, as in the hydrolysis of colloids and emulsions.

7. Changes in the reaction measured, as when a lipase catalyses the successive hydrolysis of tripalmitin, dipalmitin, and monopalmitin.

8. Changes in the molecular concentration of H_2O and other effects found in strong solutions. These are generally marked when enzymes are being used for synthesis.

Now in many cases a number of these factors are operating simultaneously. Thus in the hydrolysis of an oil emulsion by a lipase such as that of castor-oil beans which is destroyed by water all except perhaps (3) are possible, and several occur. In peptic digestion of solid proteins (1), (2), (5), (6), and (7) are all prominent. It therefore appears futile to give any but an empirical equation in such cases.

Occasionally, however, a reaction is found to follow the equation $kt = \log \frac{a}{a - x}$ of a unimolecular reaction with great accuracy. In such a case, especially where processes (5), (6), (7), and (8) can be ruled out, as in disaccharide hydrolyses, it is very likely that the reaction is proceeding according to Henri's equation, and the substrate concentration is well below the Michaelis constant of the enzyme. Such a conclusion is not, however, certain. Thus, if in the equation for sucrose hydrolysis (p. 76) we suppose that one product of the reaction has the same affinity as the substrate for the enzyme, i.e., put K_g infinite, $K_f = K_m$, $\therefore \frac{dy}{dt} = \frac{V(a - y)}{a + K_m}$. Hence the course of the reaction is unimolecular, though the initial velocity is not proportional to the substrate concentration.

To sum up, the nature of the reaction is better deduced from several rough curves representing its course under different conditions than by a single accurate curve representing it under only one set of conditions.

CHAPTER VI.

SPECIFICITY.

By specificity is meant the manner in which catalysis by an enzyme is influenced by the chemical structure of its substrates or possible substrates. We may compare either different substrates, or different linkages in the same substrate. For example, saccharase hydrolyses sucrose but not maltose, and trypsin hydrolyses some, but not all, of the peptide linkages in a protein. It may perhaps be legitimate to extend the term so as to cover the difference between the ways in which different enzymes attack the same substrate. Thus yeast zymase breaks up glucose into alcohol and carbon dioxide, muscle "zymase" into lactic acid; pancreatic amylase hydrolyses starch into α-maltose, malt amylase into β-maltose. In this chapter, however, we shall mainly confine ourselves to the first type of specificity.

Criteria for the Unity of an Enzyme.

When the facts of specificity were first discovered there was a tendency among some workers to postulate a separate enzyme for every substrate. Others went too far in the other direction. For example, it was believed that the same enzyme in emulsin hydrolysed β-glucosides and β-galactosides. The following criteria for the identity of the enzymes acting on several substrates have since been employed :—

1. They cannot be separated by fractional precipitation and adsorption. Thus yeast saccharase can readily be separated from maltase, but not from raffinase, with which it is apparently identical.

2. When one is destroyed by heat or chemical reagents so is the other. Thus the lactase (β-galactosidase) of emulsin is readily destroyed at a temperature of 45°, according to Armstrong, Armstrong and Horton [1908], and by trypsin or dialysis according to Ohta [1914]. None of these procedures destroy the β-glucosidase. This criterion must of course be applied with care where a coenzyme is present. Thus if papain-HCN is heated, the HCN is driven off, thus lowering the specificity of the papain.

3. When acting on non-electrolytes they give the same pH-activity curve, when the substrates are present in optimal concentrations. This latter condition is necessary, since K_m may vary with pH.

4. They do not give a purely additive effect when acting on a mixture of the two substrates except in low concentrations, when the enzyme is not nearly saturated. Fig. 27 from Northrop [1922, 4] shows this phenomenon in the case of trypsin acting on gelatin and casein.

5. A substance chemically related to the substrates has the same affinities (competitive or non-competitive) for both. If the affinity is competitive its addition will of course have a greater inhibitory effect on the reaction with that substrate for which the enzyme has a weaker affinity.

6. The values of K_m and V for the two substrates exhibit simple

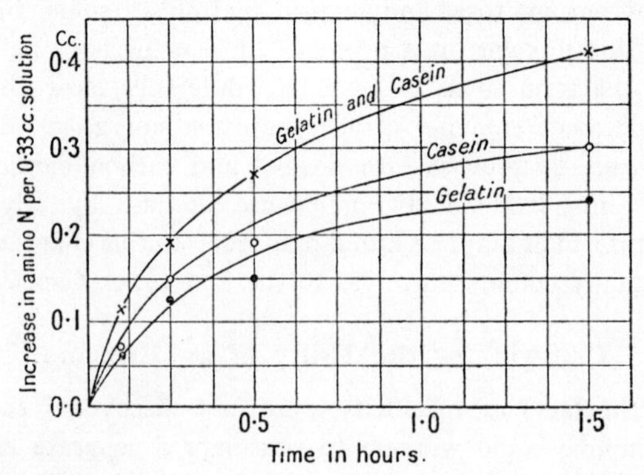

FIG. 27.—Digestion curves for 4 per cent. casein, 3 per cent. gelatin, and a mixture containing both 3 per cent. gelatin and 4 per cent. casein with the same concentration of trypsin. [Northrop, 1922, 4.]

relationships when enzymes from different sources are compared (see Chap. III.). This is not a necessary condition.

7. They are affected in the same way by coenzymes or activators. This is not always so, but is strong evidence of identity if it occurs.

Specificity among the Hexosidases.

The hexosides and pentosides present enormous possibilities of isomerism, even when the constituent sugars are known. Thus there are probably six methyl-*d*-glucosides, the α and β forms derived from (1, 4), (1, 5), and (1, 6) glucose, and four more may possibly exist. In a disaccharide the reducing group of one sugar may be attached to any [1] of the alcoholic groups of the other, or to its reducing group.

[1] Certainly to 4 or 6, probably to 3 and 5, conceivably to 2.

The specificity of an enzyme which hydrolyses a disaccharide may be influenced by either of the two sugars concerned. Thus maltose (glucose-4-α-glucoside) is hydrolysed by at least two different types of enzyme. Yeast maltase will also hydrolyse α-methyl and α-phenyl glucoside (see Chap. III.) and maltosazone according to Neuberg and Saneyoshi [1911]. On the other hand, the maltases of takadiastase (*Aspergillus*) and malt will not attack α-methyl glucoside or maltosazone, according to Leibowitz and Mechlinski [1926]. Further taka-maltase attacks α-maltose, but not β-maltose or β-methyl-maltoside [Leibowitz, 1925].

Again turanose (probably 6-α-glucosido-(2, 5) fructose) is hydrolysed by maltase-containing yeast extracts, but only slowly if at all by *Aspergillus niger* extract [Bridel and Aagard, 1927]. Hence the actual configuration of the reducing glucose residue in maltose determines its reactivity with this enzyme. It might be thought that just as yeast maltase will hydrolyse a variety of α (1, 5) glucosides, so malt maltase would hydrolyse a variety of (1, 5) glucoses substituted in the 4 position (including cellobiose). Now malt extract contains a cellobiase, which does not apparently attack β-glucosides in general, and has much the same pH optimum as malt maltase. Pringsheim and Leibowitz [1923] have shown that this enzyme can be obtained free from maltase, so the maltase is not specific for 4-hexosido-glucoses. It is not known whether malt contains a lactase. *Aspergillus* extract hydrolyses cellobiose, though very slowly [Fischer and Zemplen, 1910], and lactose only if grown on it [Pottevin, 1903]. There is thus no reason to believe that an enzyme can be specific only for the reducing half of a biose. In general it appears that enzyme specificity is primarily determined by the hexoside residue, secondarily by the attached group.

The lactases appear to differ among themselves like the maltases. Thus Bierry and Giaja [1908] and Bierry [1909] found that Helix intestinal juice will hydrolyse lactosazone and lactobionic acid, while mammalian gut lactase will not attack lactobionic acid, and according to Neuberg and Saneyoshi [1911] the lactases of emulsin and kefir only attack lactosazone extremely slowly. But Fischer and Curme [1914] found that almond lactase can hydrolyse lactal and hydrolactal. Similarly Zemplen [1921] found that it can hydrolyse 3 (1, 5) β-galactosido-*d*-arabinose, but not 2 (1, 5) β-galactosido-erythrose. Since kefir lactase will synthesize an isolactose (? 6-β-galactosido glucose) [Fischer and Armstrong, 1902], and almond lactase two galactose galactosides [Bourquelot and Aubry, 1917], it is clear that the position of attachment of the galactoside residue to the reducing sugar is not relevant to these enzymes. They can, moreover, synthesize a variety of alkyl-, and also salicyl-β-galactosides [Bourquelot, 1917].

These different types of specificity become more intelligible if the reverse actions are considered. A maltase molecule can presumably unite with one of α-glucose, which is so modified as to be particularly prone to form a glucoside, and also with a second molecule. In the case of yeast maltase this latter may be almost any alcohol; in that of *Aspergillus* maltase, it must be α-glucose. Yeast maltase as a synthesizing enzyme is singly specific; *Aspergillus* maltase doubly specific.

Specificity of the β-glucosidase of Emulsin (Prunase).

Fischer [1919] and others have found that emulsin will hydrolyse a large number of β-glucosides. Glucosides of a number of monohydric and polyhydric alcohols and phenols were split, as were those of phenol-alcohols, phenol-aldehydes, and phenol-ketones, and the β-glucoside sugars primeverose, gentiobiose, and cellobiose, with cellosidoglycollic acid and its amide and nitrile, hydrocellobial and cellobiosone, and even methyl and ethyl thiocellobiosides [Wrede and Hettche, 1927]. The work of Helferich, Löwa, Nippe, and Riedel [1923], who found that emulsin has two pH optima (from Cu reduction) when acting on methyl cellobioside, suggests the possibility of a specific cellobiase or cellobiosidase in addition to β-glucosidase. A Br atom in the alcoholic residue, as in bromallyl-glucoside had no inhibitory effect. The glucosides of acids were also in general hydrolysed, but the behaviour of the derivatives of two of them, namely, α-hydroxy-isobutyric and mandelic acids, were peculiar. The salts of β-glucosido-α-hydroxy-isobutyric acid were barely, if at all, hydrolysed, its methyl-ester, amide, and nitrile very slowly. The *d*-mandelic acid glucoside and its amide were practically untouched, whereas the *l*-mandelic amide glucoside was easily hydrolysed. On the other hand, both the *d*- and *l*-nitriles, and the methyl ester of glucosido-*d*-mandelic acid were affected. The resistance of the glucosides of these acids, which are tertiary and secondary alcohols, agrees with the very slow hydrolysis of tertiary amyl alcohol β-glucoside. Of course, however, the natural substrate of emulsin β-glucosidase is prunasin (*d*-mandelo-nitrile [1] β-glucoside). This is hydrolysed al-

[1] The nitrile of *l*-mandelic acid.

though *d*-mandelonitrile is a secondary alcohol, and so is sambunigrin (*l*-mandelonitrile β-glucoside). But configuration has a very large effect on the rate of hydrolysis of these secondary alcohol glucosides, as is shown by the different behaviour of the glucosides of *d*- and *l*-mandelamide, and of those of *d*-mandelic acid and its methyl-ester. It is noteworthy that the amide of the natural product of hydrolysis, *l*-mandelic acid, is the isomer acted upon. As β-glucosidase has been shown (see Chap. III.) to have an affinity for alcohols and their glucosides, it is possible that, in the case of its natural substrate, this may outweigh the effect of the steric hindrance due to the secondary character of the alcoholic residue. Indeed the affinity of emulsin for amygdalin is very high.

Besides representatives of the bioses, and glucosides of aliphatic and aromatic alcohols analogous to those hydrolysed (Fischer), Bourquelot [1914] and his colleagues have synthesized β-glucosides of cetyl alcohol, cyclohexanol, *o*-methyl-cyclohexanol, borneol, and morphine with emulsin. It is not, of course, certain that all these substances are formed by prunase, but there is no serious reason to doubt it.

The Relative Specificities of α- and β-Glucosidase.

Yeast maltase and emulsin respectively hydrolyse a number of α- and β-glucosides, but the former shows a far greater specificity as regards the hexoside part of the molecule. Reduction of the sixth carbon atom of β-methyl glucoside yields β-methyl-*d*-epirhamnoside, which is split by emulsin [Fischer, 1919], while the correspondingly reduced α-methyl *d*-epirhamnoside is not affected by maltase [Helferich, Klein, and Schäfer, 1926]. Similarly, the addition of another glucose residue inhibits maltase but not emulsin. The former does not attack α-methyl gentiobioside [Helferich and Becker, 1924], the latter hydrolyses β-methyl maltoside into maltose and methyl alcohol.

While so far yeast maltase has not been found to attack any α-glucosides in which the α-glucose residue has been altered, emulsin behaves to β-hexosides and pentosides according to Table XIV.

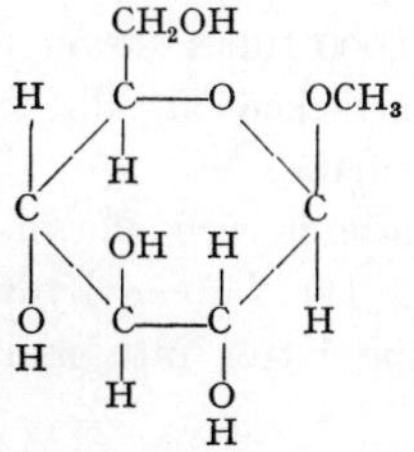

```
        CH2OH                              CH3
         |                                  |
 H       C——O     OCH3             H        C——O     OCH3
 |     / |    \    |               |      / |    \    |
 |   /   H      \  |               |    /   H      \  |
 C                 C               C                  C
 | \  OH   H    /  |               | \   OH   H    /  |
 |   \ |   |  /    |               |   \  |   |  /    |
 O     C———C       H               O      C———C       H
 H     |   |                       H      |   |
       H   O                              H   OH
           H
```

β-methyl (1, 5) *d*-glucoside. β-methyl (1, 5) *d*-epirhamnoside.

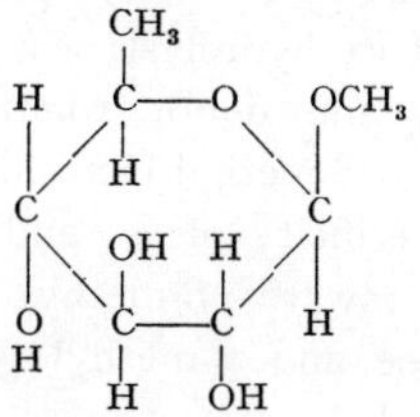

```
         CH2Br                                  H
          |                                     |
  H       C——O      OCH3                H       C——O      OCH3
  |     / |    \     |                  |     / |    \     |
  |   /   H      \   |                  |   /   H      \   |
  C                  C                  C                  C
  | \    OH   H    / |                  | \    OH   H    / |
  |   \   |   |  /                      |   \   |   |  /
  O       C———C      H                  O       C———C      H
  H       |   |                         H       |   |
          H   OH                                H   OH
```

β-methyl (1, 5) *d*-glucoside 6-bromhydrin. β-methyl (1, 5) *d*-xyloside.

TABLE XIV.

Substrate.	Hydrolysis.	
β-*d*-glucosides	+	
β-methyl--*d*epirhamnoside	+	Fischer (1919).
β-methyl-*d*-xyloside	—	Fischer (1895).
β-methyl-*d*-glucoside 6-bromhydrin	—	Fischer (1919).
β-methyl-gentiobioside	+	Helferich and Becker (1924).
β-methyl-*l*-glucoside	—	Fischer (1894).
β-methyl-*d*-mannoside	—	Fischer (1894).
β-methyl-glucoside monophosphate and diphosphate	—	Helferich, Löwa, Nippe, and Riedel (1923).
β-methyl-glucoside monosulphate	—	Helferich, Löwa, Nippe, and Riedel (1923).
Saligenin β-benzoyl-glucoside	—	Kitasato (1927).

The hydrolysis of β-galactosides is certainly, that of β-ethyl *l*-arabinoside probably, due to a different enzyme, as is possibly that of the primeverosides. A hexose residue may therefore be substituted for an alcoholic H atom. Benzoylation of the glucose inhibits the β-glucosidase of almonds, but not of takadiastase [Kitasato, 1927].

Methyl can be substituted for the 6-methoxyl group, but not hydrogen for the methyl. And Br cannot be substituted for OH in this position, though it can be substituted in the alcoholic part of the glucoside molecule. No change is permissible in the position of the O atoms attached to the carbon atoms of the amylene oxide ring of the molecule, to which presumably the enzyme is specifically adapted.

Purdie and Irvine [1904] state that tetramethyl-β-methyl-glucoside is hydrolysed by emulsin, but Kuhn and Schlubach [1925] found that the rate of its hydrolysis was at least 60,000 times slower than that of helicin, and doubt whether it is attacked at all. Similarly, heptamethyl-β-methyl-lactoside is unaffected.

The specificity of α- and β-glucosidase is well shown by their behaviour towards β-methyl maltoside. The former hydrolyses it into glucose and β-methyl glucoside, the latter into maltose and methyl alcohol.

Specificity of the Invertases (Saccharases).

The various enzymes which hydrolyse saccharose may be further differentiated by their behaviour towards raffinose, melezitose, gentianose, and "hesperonal," which are substituted saccharoses. Raffinose is saccharose β-galactoside, the galactose residue being attached to the glucose residue in the 6-position, so that its hydrolysis yields saccharose and galactose, or fructose and melibiose. Melezitose is a sucrose glucoside in which the extra glucose is attached to the fructose residue. It can be hydrolysed into glucose and turanose (6-glucosido-fructose). Since this is hydrolysed by bottom yeast extract but not by emulsin, it is believed by Bridel and Aagard [1927] to be an α-glucoside. These authors, however, do not regard the glucosido-fructoside residue as sucrose. Gentianose is sucrose β-glucoside, the β-glucose being attached to the fourth carbon atom of the glucose residue. It can thus be hydrolysed into fructose and gentiobiose. Thus, if < represents the reducing group we may probably write :—

Raffinose = $\overbrace{\text{Fructose} < > \text{glucose}}^{\text{Sucrose}}$ > galactose, where glucose > galactose = Melibiose

Melezitose = $\overbrace{\text{glucose} < \text{fructose}}^{\text{Turanose}}$ < > glucose, where fructose < > glucose = Sucrose

Gentianose = $\overbrace{\text{glucose} < \text{glucose}}^{\text{Gentiobiose}}$ < > fructose, where glucose < > fructose = Sucrose

Hesperonal is the synthetic saccharose monophosphoric acid of Neuberg and Pollak [1910], the phosphoric acid being attached to the glucose residue, since it yields glucose monophosphoric acid and fructose on hydrolysis.

Highly purified yeast invertase hydrolyses raffinose into fructose and melibiose, hesperonal into fructose and glucose monophosphoric acid [Kuhn and Münch, 1925]. It also hydrolyses β (1, 5) methyl fructoside (one of the components of the so-called *h*-methyl fructoside) to fructose and methyl alcohol [Schlubach and Rauchalles, 1925]. It does not, however, attack melezitose in which the fructose residue is substituted. Crude yeast invertase hydrolyses gentianose to gentiobiose and fructose [Bourquelot and Bridel, 1911]. It is clear, therefore, that yeast invertase is specific for β (1, 5) fructosides. It is described as a fructosaccharase. Some fructosans, e.g. synanthrine and graminine are attacked by yeast saccharase preparations, possibly by their saccharase [Colin and Cugnac, 1926]. The majority are not so attacked.

Other invertases have not been so highly purified, but invertase-containing solutions from *Aspergillus* and other moulds and the mammalian gut have been extensively studied. Kuhn and Grundherr [1926] found that extracts of *Aspergillus niger* and *Penicillium glaucum* hydrolyse melezitose to glucose and turanose, while they do not affect raffinose, and *Aspergillus* extract hydrolyses gentianose to glucose and sucrose before attacking the latter [Bourquelot, 1898]. Kuhn and Münch [1925] found that *Aspergillus oryzae* extract hydrolysed hesperonal to glucose, fructose, and phosphoric acid, but as the number of free phosphoric acid molecules was always slightly greater than that of free glucose or fructose molecules, it appeared that the phosphoric acid was first removed by a phosphatase, and the sucrose so formed subsequently hydrolysed. This agrees with the results of Helferich, Löwa, Nippe, and Riedel [1923], who found that the introduction of a sulphuryl or phosphoric acid residue prevented hydrolysis of the alkyl and phenyl-derivatives of α-glucose, β-glucose, and cellobiose, but not of maltose or amygdalinic acid, where the non-glucosidic sugar residue may have been substituted. *Aspergillus* invertase therefore requires that the glucose residue of sucrose should be unaltered, and is consequently called a glucosaccharase. These conclusions are contested by Bridel and Aagard [1927], who found that preparations of *Aspergillus niger* having equal activities on sucrose possessed very different activities when tested on melezitose. They therefore believe that melezitase and invertase are different enzymes. It is, however, conceivable that the differences may be due to variations in affinity analogous to those found in different invertase-raffinase preparations from yeast. Moreover, even *Aspergillus oryzae* may yield preparations having the properties of a fructosaccharase [Rohdewald, 1927].

Mammalian gut invertase does not hydrolyse raffinose, but its behaviour to hesperonal and melezitose is unknown. It is generally thought to be a glucosaccharase. On the other hand, some invertebrate intestinal juices hydrolyse raffinose.

Since fructosaccharase hydrolyses the corresponding methyl fructoside, it would be expected that glucosaccharase would attack α-methyl glucoside and other α-glucosides, and in fact be identical with a maltase of the yeast type. Weidenhagen [1928] found that yeast maltase, freed from the ordinary saccharase, hydrolyses saccharose at about the same rate as maltose at pH 7·0, which is near the optimum for maltase action, but not at pH 4·6, the optimum for saccharase (β (2, 5) fructosidase). If this remarkable observation is confirmed it is clear that the term "saccharase" will become obsolete, and that saccharases will ultimately be classified according to the hexose residue for which they are specific.

Similarly, mammalian gut maltase and saccharase have the same optimum pH, and may be identical, while it is also possible that mammalian gut saccharase is responsible for the inversion of trehalose. As against this view Fischer and Niebel [1896] found maltase present in all vertebrate sera and tissues investigated by them (though it is absent in others). Trehalase was only present in some, and invertase still rarer. Either, therefore, the enzymes are different, or the α-glucosidases differ in their specificity.

Inhibition and Specificity.

Different hexosidases differ in regard to the sugars which inhibit their action. Thus yeast saccharase is always inhibited by fructose and β-glucose, and often not by α-glucose, whereas *Aspergillus oryzae* (takadiastase) saccharase is not inhibited by β-glucose and fructose [Kuhn and Münch, 1925] but is strongly inhibited by α-glucose. *Penicillium glaucum* saccharase is inhibited by glucose but not fructose, while the relative influence of α- and β-glucose is not known. Mammalian gut saccharase is more strongly inhibited by glucose than by fructose [Cajori, 1930], as gut lactase, in contrast to that of emulsin, is more inhibited by glucose than galactose [Stephenson, 1912]. This is usually taken to show that yeast saccharase is adapted to the fructose part of the saccharose molecule, that of moulds to the glucose part, a result borne out by experience with the substituted sucroses. But since the fructose residue of saccharose is (2, 5) fructose, and the inhibiting sugar is (2, 6) fructose, this fact must, it would seem, be regarded as accidental. Such inhibition is an inadequate guide to specificity, as appears from the fact that, for example, β-glucose and galactose inhibit yeast invertase. It is valuable as distinguishing between otherwise similar enzymes (e.g. the yeast saccharases which are and are not inhibited by α-glucose), and in establishing the identity of an enzyme acting on different substrates (e.g. xanthine oxidase).

Specificity of the Lipases.

The possible substrates for lipase are to be numbered in millions. A huge variety of esters can be hydrolysed. The acids range from formic to the higher fatty acids in the aliphatic series, and from benzoic to morphine-glycollic [Schmidt, 1901] in the aromatic. Esters of dibasic acids may be hydrolysed completely or incompletely. Halogen and amino groups or branched chains do not prevent the action. It is not yet clear whether the tannase of mammalian tissues [Sieburg and Mordhorst, 1919] is the same as lipase. Less is known as to the range of alcohols whose esters can be hydrolysed. Ordinary lipases hydrolyse esters of the aliphatic alcohols from methanol to the various pentanols, and it is very probable that the hydrolysis of cetyl, ceryl, myricyl, and cholesteryl esters by mammalian organs [Porter, 1916] is due to the same substances, more especially as the activity of a tissue on waxes runs parallel with that on the higher fats rather than on soluble esters. All lipases attack the esters of glycerol, and even those of its chlorhydrins are hydrolysable [Abderhalden and Weil, 1920]. It is not clear whether the enzymes in mammalian tissues hydrolysing phenyl salicylate [Nencki, 1886], acetyl choline [Loewi and Navratil, 1926], and atropine [la Barre, 1925] are to be identified with ordinary lipase, nor whether purified lipase is capable of a partial hydrolysis of lecithin. On the other hand, chlorophyllase appears to be quite distinct from the lipases [Willstätter and Stoll, 1913], and lecithin hydrolysis, as determined by acidimetry, is probably often largely due to a phosphatase. Bloor [1912] found that mannide distearate was hydrolysed by pancreatic and castor oil lipases, while mannitan distearate (which contains two more hydroxyl groups) was digested by the former, but very slowly if at all by the latter. Other authors have found that various esters of mannitol anhydrides are easily digested.

The possibility remains that all lipase preparations so far obtained are mixtures, and may in future be separated into components of greater specificity. Nevertheless, the purified preparations of lipase from stomach, pancreas, liver, and seeds, obtained by Willstätter and his pupils, can hydrolyse esters of volatile and non-volatile aliphatic, as well as of aromatic acids, of simple monohydric alcohols and of glycerol. They have, unfortunately, not yet been tested on depsides, waxes, or cholesteryl or phenyl esters. It may, therefore, be assumed provisionally that the specificity shown by them is quantitative and not absolute.

Whereas the lipases of stomach, pancreas [Willstätter and Memmen, 1924], and *Ricinus* seeds [Armstrong and Ormerod, 1906] are more efficient in the hydrolysis of fats than of the mono-esters of the lower fatty acids, the opposite is true of liver and *Aspergillus* lipases, which may better be termed esterases. The difference is at least in part due to the relatively slight affinity of the true lipases for the lower esters (see Chap. III.). It is very large. Thus, whereas a certain quantity of pancreas was equally active with a certain amount of liver in the hydrolysis of methyl butyrate, it was 250 times as active on tributyrin, and 26,000 times as active on olive oil [Willstätter and Memmen, 1924]. These differences may be due in part to coenzymes, but can hardly be so entirely.

The lipases exhibit another type of quantitative specificity. They act at different rates on optical antipodes, whether the asymmetry is due to the acid radical [Dakin, 1903] or the glycerol residue, as in the cases of the hydrolyses of lecithin [Mayer, 1906], monacetin and monocaproin [Abderhalden and Weil, 1920]. The first type of specificity has been mainly investigated, and the principal results are given in Table XV.

TABLE XV.

HYDROLYSIS OF RACEMIC ESTERS.

Substrate.	Pancreas.	Liver.	Stomach.	Aspergillus.
Ethyl lactate	—	L	—	—
Diethyl and isobutyl tartrates . .	L	D	—	—
Ethyl methylethylacetate . .	D	—	—	—
,, methylbutylacetate . .	D	—	—	—
,, mandelate	L	D	D	D
Methyl ,, ...	L	D	—	D
Glyceryl monomandelate . .	L	D	—	—
Methyl phenylmethoxyacetate . .	L	D	—	D
,, phenylchloracetate . .	L	L	D	L
Ethyl phenylchloracetate . .	—	L	—	—
Methyl phenylbromacetate . .	L	L	—	—
Propyl phenylaminoacetate . .	D	D	—	—
Ethyl methylbenzylacetate . .	D	—	—	—
Propyl leucinate	D *	D *	—	—
Ethyl leucinate	D *	—	—	—
,, tyrosinate	L	—	—	—
Methyl tropate	D	L	D	L

From Dawson, Platt, and Cohen (1926), Waldschmidt-Leitz (1926), Rona and Ammon (1927), and Rona and Itelson-Schechter (1928).

D denotes preferential hydrolysis of the dextro-rotatory ester (not necessarily the ester of the dextro-rotatory acid), and L the converse

* Esters of *l*-leucine.

in a racemic mixture. In addition no appreciable specificity is shown between the *d*- and *l*-components of certain *r*-esters. It will be seen that pancreas and stomach lipases, which agree roughly in their quantitative specificity as between fats and esters, sometimes differ in this respect. What is more, as they agree with regard to their action on ethyl mandelate, they cannot be optical antipodes. According to Willstätter, Bamann, and Waldschmidt-Graser [1928], the differences between different lipases in this respect are not altered by asymmetrical activators such as proteins, and are therefore probably due to real difference in the enzyme molecule. It is worthy of note that in all cases the optical activity of the acids is due to an asymmetrical α C atom. We do not know whether most lipases are sensitive to an asymmetry more remote from the ester linkage. But Neuberg and Rosenberg [1907] obtained *d*-dibrom stearic acid from triolein hexabromide with Ricinus lipase.

Rona and Ammon [1927] have shown that human liver differs from the majority of mammalian livers in possessing a lipase which hydrolyses *l*-methyl mandelate preferentially. Guinea-pigs' blood has the same effect. It might be supposed that an enzyme which splits a racemic ester preferentially would act more rapidly on one component than the other when alone. This is not necessarily the case. Rona and Ammon found that *Aspergillus* lipase, which hydrolyses methyl *d*-mandelate preferentially in a racemic mixture, hydrolyses the *d*-, *l*-, and *dl*-esters at exactly the same velocity. While pig's pancreatic lipase, as might be expected from its behaviour with the *dl*-ester, hydrolyses the *l*-, *dl*-, and *d*-esters in that order of velocities, pig's liver lipase, although it hydrolyses the *d*-ester preferentially in a mixture, hydrolyses the *d*- and *dl*-esters at an equal velocity, and the *l* considerably faster than either. Now all these substrates were used in concentrations far above the saturation limit of the enzyme, and the results can be explained as follows :—

Aspergillus lipase has a higher affinity for the *d*- than the *l*-ester, but both are split at the same velocity when the enzyme is saturated. Pig's liver lipase has a much higher affinity for the *d*- than the *l*-ester, but hydrolyses the latter more rapidly when saturated with it. This explanation is fully borne out by the work of Willstätter, Kuhn, and Bamann [1928] and Bamann [1929] on liver lipases. Thus pig's liver lipase has the following constants for the ethyl mandelates :—

Ethyl *d*-mandelate . . . . $V = 1$ $K_m = \cdot 0005$.
,, *l*- ,, ... $V = 1{\cdot}5$ $K_m = \cdot 0016$.

I.e. if the velocity of hydrolysis is plotted against substrate concentration for the two esters, the curves cross. The course of events in this case is considered in Chapter V., pages 85-88.

Specificity of Phosphatases.

Kay [1928] produces strong evidence for the identity of the enzymes in all mammalian tissues which hydrolyse glycerophosphoric acid, fructose diphosphoric acid, and nucleotides.

Mammalian and *Aspergillus* phosphatases have a very wide range of specificity [Neuberg and Wagner, 1926, 1; Iwatsuru, 1926; Neuberg and Jacobsohn, 1928]. Mono-orthophosphoric esters of the following radicals, among others, are hydrolysed :—

Methyl, ethyl, amyl, cetyl, cholesteryl, phenyl, bornyl, α- and β-glyceryl, glucosyl (2 isomers), sucrosyl.

In addition diphenyl orthophosphoric ester, and (2, 5) fructose diorthophosphoric ester are attacked.

The same extracts cause hydrolysis of pyrophosphates [Lohmann, 1928], and the following di-pyrophosphoric esters :—

Phenyl, *o*-cresyl, *m*-cresyl, α-naphthyl.

In the latter three cases no free pyrophosphoric acid appears as an intermediate. It is not yet clear whether in the case of the pyrophosphates a single enzyme is concerned, or whether a special pyrophosphatase first attacks the pyrophosphate grouping, and the resulting orthophosphoric ester is then hydrolysed by ordinary phosphatase. Metaphosphates are also hydrolysed by a variety of tissue extracts [Kitasato, 1928]. Samec and Knez [1925] find that a variety of extracts, including that of bone, attack amylopectin and synthetic amylophosphoric acid, and ascribe this action to the ordinary phosphatases.

There are, however, limits to the range of specificity of these phosphatases. Tissue phosphatase does not attack casein, but removes either two or all of the three phosphate radicles from Rimington's enneapeptide triphosphoric ester, a casein derivative [Rimington 1927]. No phosphatase has so far been found to attack trimethyl phosphate [Neuberg and Jacobsohn, 1928], and bone phosphatase only attacks dimethyl phosphate very slowly [Martland and Robison, 1927]. Moreover, while hexosephosphoric esters and their methyl hexosides are hydrolysed, this capacity is lost on complete methylation. The kidney phosphatase of rabbits and certain other mammals (not including the pig) has a wider range of action than that of bone and other tissues. It is the only enzyme so far known to hydrolyse the third phosphate radical of phosphopeptone [Rimington, 1927, 2], or to attack diphosphoglyceric acid [Kay, 1926; Jost, 1927]. Phosphatases show a quantitative specificity with regard to the esters of the optical isomers *dl*-borneol and *dl*-methyl-propyl-carbinol [Neuberg, Wagner and Jacobsohn, 1927; Neuberg and Jacobsohn, 1928], the *l*-alcohol being first liberated. Further details of specificity as revealed by arsenate activation are given in Chapter VII.

Specificity of Sulphatase, Urease, Hippuricase, and Asparaginase.

Sulphatase attacks a large variety of aromatic sulphuric esters, including those of simple phenols and naphthols and their homologues, diphenols, phenolaldehydes and phenol-carboxylic acids, and also of oxyquinoline and indoxyl. But it does not attack the ethereal sulphates of alkyls, *m*-methyl-cyclohexanol, borneol, or even mandelic acid [Neuberg and Wagner, 1925]. Nor are carbohydrate esters or sinigrin attacked, while myrosinase, the enzyme attacking the latter, has no action on aromatic sulphates [Neuberg and Wagner, 1926, 2].

Ureases are generally very specific, but Pin Yin Yi [1920] claims that *Robinia* urease will also attack asymmetrical dimethyl- and diethyl-ureas. Preparations of hippuricase (or histozym) hydrolyse not only benzoyl-glycine but also its homologues and related bodies such as benzoyl-alanine [Mutch, 1912], pyromucuric acid, acetyl-glycine, butyryl-alanine, butyryl-glycine, lauryl-glycine, lauryl-alanine, and lauryl-alanyl-glycine and also glycocholic and taurocholic acids [Smorodinzew, 1923]. As preparations having the above properties did not hydrolyse glyceryl tribenzoate [Nencki, 1886], nor does erepsin hydrolyse hippuric acid [Cohnheim, 1907], it seemed reasonable to attribute the hydrolyses of the hippuric acid homologues to the same catalyst. However, Kimura [1928, 1929] showed that hippuricase has a more restricted distribution in the body than the enzyme hydrolysing acetyl-glycine, formyl-glycine, and homologues such as acetyl-leucine. There is evidence that a third distinct enzyme hydrolyses acetyl-phenylalanine. Smorodinzew [1923] further showed that hippuricase does not attack benzoylated β-amino-acids such as benzoyl-β-alanine and benzoyl-β-amino-butyric acid, nor benzoyl-amino-isobutyric acid. It also possesses stereochemical specificity, hydrolysing benzoyl *d*-alanine, but benzoyl-*l*-alanine slowly or not at all [Mutch, 1912], and *d*-benzoylaminobutyric acid, but not its antipode [Smorodinzew, 1923]. To sum up, it will hydrolyse

$$R—CH_2—CO—NH—CH_2—COOH,$$

one component of

$$R—CH_2—CO—NH—\overset{R'}{\underset{H}{C}}—COOH$$

but not

$$R—CH—CO—NH—\overset{CH_3}{\underset{CH_3}{C}}—COOH.$$

Geddes and Hunter [1928] found that yeast asparaginase (a deaminating enzyme) attacked asparagine and glutamine but not urea, formamide, acetamide, valerianamide, oxamide, succinamide, or salicylamide.

Specificity of Arginase.

This enzyme, which hydrolyses *d*-arginine to ornithine and urea, has been tested on guanidine [Clementi, 1916], guanidine-acetic, β-guanidine-propionic [Edlbacher, 1917], γ-guanidine-butyric and ϵ-guanidine-capronic acids [Thomas, 1913]; and also creatine. Of these only γ-guanidine butyric acid was affected. *l*-arginine is also unaffected [Riesser, 1906], while arginyl methyl ester is hydrolysed slowly if at all [Edlbacher and Bonem, 1925]. Agmatine is unaffected by mammalian or fungal arginase, but Kiesel [1922] found that an arginase preparation of *Aspergillus niger* could split tetramethylene diguanidine (α-, δ-diguanidyl butane) into agmatine and urea. Arginine residues are not attacked in the protein molecule, though the guanidine group is free [Kossel and Dakin, 1906]. But Edlbacher and Bonem [1925] found that half the guanidine groups of arginylarginine are hydrolysed, and believe that the hydrolysis occurs in the arginine residue with free carboxyl group. These results are somewhat surprising, especially the failure to attack δ-guanidine-caproic acid (desamino-arginine) while γ-guanidine-butyric acid is attacked.

The Enzymes Hydrolysing Proteins and Peptides.

Table XVI. gives a summary of the ranges of specificity among natural products of those enzymes of this group which have been investigated in a purified form.

TABLE XVI.

Enzyme.	Native Albumin.	Denatured Albumin, Gelatin, etc.	Histones.	Protamines.	"Peptone" (Peptic Digest).	Polypeptides.	Dipeptides.	Prolyl Peptides.
Dipeptidase [1, 6]	−	−	−	−	−	−	+	−
Amino polypeptidase [1, 6]	−	−	−	−	±	+	−	−
Prolylpeptidase [2]	−	−	−	−	−	−	−	+
Carboxy-polypeptidase [7]	−	−	−	+	+	+	+	−
Trypsinogen [7]	−	−	−	−	−	−	−	−
,, +kinase [3, 7]	+	+	+	+	?	−	−	−
Kathepsin [4]	−	−	−	−	?	−	−	?
,, + activator [4]	+	+	?	+	?	−	−	?
Yeast protease [1]	−	−	−	−	+	−	−	−
,, ,, +HCN [1]	+	+	+	+	++	−	−	−
Papain [5]	−	+	−	−	±	−	−	−
,, + HCN	+	++	+	+	+	+	−	−
Pepsin	+	+	+	+	−	−	−	−

The sign + denotes that some members of the group are hydrolysed. But it must be added that, for example, whereas dipeptidase hydrolyses a great many dipeptides, carboxy-polypeptidase from pancreas only attacks a few. The sign + + denotes increased activity on adding an activator, the sign − that no effect has so far been observed. ± denotes a very slight effect. ? denotes inadequate information. Kathepsin without activation is said to digest "certain protein breakdown products."

The first three enzymes on the list are the probable components of erepsin. Crude erepsin preparations from gut or yeast digest a variety of dipeptides and polypeptides, including those in which the free amino group is not, as in most peptides, primary, but the secondary amino group of proline.

```
CH2—CH2
|      \
CH2     CH—CO—NH—R—COOH.
  \    /
    NH
```

The yeast erepsin has been fractionated into a polypeptidase and a dipeptidase, gut erepsin has also been fractionated, but it is merely

[1] Grassmann and Dyckerhoff (1928, 2).
[2] Grassmann, Dyckerhoff, and Schoenebeck (1929).
[3] Waldschmidt-Leitz (1926).
[4] Waldschmidt-Leitz, Bek, and Kahn (1929).
[5] Willstätter and Grassmann (1924).
[6] Waldschmidt-Leitz, Balls, and Waldschmidt-Graser (1929).
[7] Waldschmidt-Leitz and Purr (1929).

known that one component hydrolyses leucyl-glycine specifically, the other leucyl-glycyl-glycine. It is assumed for the sake of simplicity that as in the case of yeast, the boundary between their fields of action is that between dipeptides and polypeptides. The dipeptidase also appears to be a mixture [Linderström-Lang and Sato, 1929] of an unstable enzyme (dipeptidase I) hydrolysing leucyl-glycine and glycyl-glycine optimally at pH 7·3, and a stabler enzyme (dipeptidase II) with an optimum at pH 8·1. The first acts at about equal rates on the two substrates in ·1 M solution, the second acts about twenty times more rapidly on leucyl-glycine than glycyl-glycine. The two have not been completely separated.[1] Most of the published results probably refer to mixtures consisting mainly of dipeptidase II. On purification by adsorption both yeast and gut peptidases lose the capacity for hydrolysing prolyl-glycine and prolyl-glycyl-glycine possessed by crude extracts. It is presumed that this capacity is due to a special enzyme. It may possibly, however, be due to a special activator of the peptidases. Waldschmidt-Leitz and Purr [1929] have separated "trypsin" into four fractions. At pH 4 aluminium-hydroxide C γ removes dipeptidase and amino-polypeptidase. These are the same as those of gut erepsin, and can be separated by adsorption on ferric hydroxide at pH 4, after elution from the alumina.

The residual solution can be fractionated with the same aluminium hydroxide at pH 7. This removes proteinase, leaving in solution a peptidase which attacks a number of polypeptides and some dipeptides. It does not require a free amino group in its substrate, but requires a free carboxyl (see p. 115). The proteinase is the classical trypsinogen. It does not act except in presence of enterokinase (see p. 140) and then only on large molecules. The carboxypeptidase, in many cases at least, appears to act in the absence of enterokinase; but enterokinase, or some body associated with it, increases the activity of the peptidase.

The enzymes attacking proteins, called proteinases by the Willstätter school, with the exception of pepsin, agree in being inactive or having a narrow range of specificity in the absence of activators. Both the digestive enzyme, trypsin, and the intracellular enzymes (kathepsin is the name given to the intracellular

[1] Grassmann and Klenk (1930) ascribe these results to affinity. Some dipeptidase preparations have a high affinity (K_m = ·001 M) for leucyl-glycine, and a lower (K_m = ·05 M) for glycyl-glycine. Alanine, a competitive inhibitor, depresses the hydrolysis of the latter only. Linderström-Lang's results may perhaps be explained by the presence of such inhibitors, and the variation of affinity with pH.

proteinase of mammalian organs) agree in being associated with peptidases in the cells, and their specificity only appears after purification.

There is a very wide parallelism between the enzymes of animal and plant cells. However, the proteinases differ in their specificity with regard to certain proteins, and the peptidases in their specificity on compounds other than the naturally occurring peptides. It is extremely difficult to summarize the literature, since all properties ascribed to trypsin before 1929 are due to a mixture of trypsin and peptidases, and all properties ascribed to animal erepsin before 1929 to a mixture of dipeptidase and polypeptidase.

Stereochemical Specificity of Peptidases and Proteinases.

All amino-acids but glycine have asymmetrical molecules. Hence the number of possible isomers of a peptide due to stereoisomerism is 2^n, where n is the number of amino-acid residues other than glycine. If, for example, alanyl-leucine is made synthetically from optically inactive materials all four isomers may be obtained. When α-brom-propionic acid is condensed with *dl*-leucine two racemic α-brom-propionyl-leucines of different solubilities can be separated. One of these yields the racemic alanyl-leucine A, which consists of equal parts of *d*-alanyl-*l*-leucine and *l*-alanyl-*d*-leucine; the other alanyl-leucine B, which consists of *d*-alanyl-*d*-leucine and *l*-alanyl-*l*-leucine. The latter mixture is unattacked, of the former only the *d*-alanyl-*l*-leucine moiety is attacked. This may be represented as

$$
\begin{array}{ccccccc}
 & CH_3 & & & C_4H_9 & & \\
 & | & & & | & & \\
NH_2— & C & —CO—NH— & & C & —COOH \\
 & | & & & | & & \\
 & H & & & H & &
\end{array}
$$

The same principle holds for all dipeptides and all dipeptidases (including the carboxy-peptidase associated with trypsin) so far investigated. Examples, and references to the earlier work of Abderhalden and his colleagues, are given by Levene, Steiger, and Bass [1929]. Since glycine peptides are readily hydrolysed we cannot say that asymmetry is necessary. But if, as seems probable, the configuration of groups round the active C atom is similar in all the natural optically active amino-acids, it would seem that the CH_3 and C_4H_9 groups in the above formula may be replaced by H or other radicals, but the two H atoms must probably remain intact, since Levene, Steiger, and Bass [1929] found no hydrolysis by erepsin of dipeptides with a tertiary carbon atom next to the carboxyl (e.g. glycyl α-amino-isobutyric acid).

The case of polypeptidases is more complicated. As will be seen later polypeptides appear to be attacked by erepsin at the end bearing a free amino group, by the peptidase associated with trypsin at that bearing a free carboxyl. Now consider the tripeptide *dl*-leucyl-glycyl-*l*-tyrosine. This consists of two (non-antipodal) isomers, *l*-leucyl-glycyl-*l*-tyrosine, and *d*-leucyl-glycyl-*l*-tyrosine, the former consisting entirely of natural amino-acid residues, the latter containing the natural *l*-tyrosine and the unnatural *d*-leucine. Abderhalden and Schapiro [1927] found that yeast and animal erepsins hydrolyse the natural tripeptide. As they are mixtures of dipeptidase and polypeptidase they finally hydrolyse it completely. But they do not attack *d*-leucyl-glycyl-*l*-tyrosine at all, since the hydrolysis would have to begin at the *d*-leucyl end, the enzymes being attached to the free amino group. On the other hand, Waldschmidt-Leitz and Schlatter [1928] found that trypsin preparations containing carboxy-peptidase remove tyrosine from both compounds. Hence the action of " trypsin " is not inhibited by the presence of an unnatural amino-acid residue, provided that it is not in the immediate neighbourhood of the peptide linkage undergoing hydrolysis.

Abderhalden and Handovsky [1921] found that yeast erepsin completely hydrolysed glycyl-*l*-leucine and glycyl-*l*-leucyl-glycyl-*l*-leucine, but not glycyl-*d*-leucine or glycyl-*d*-leucyl-glycyl-*l*-leucine (in three out of eight experiments there was a decrease of about half in the amino nitrogen of the latter compound, suggesting synthesis). Here again the unnatural amino-acid residue was in the neighbourhood of the free amino group. Similarly Abderhalden and Schwab [1928] found that *l*-leucyl-glycyl-*l*-leucine was hydrolysed by both " trypsin " and erepsin, *d*-leucyl-glycyl-*l*-leucine by " trypsin " only, and *l*-leucyl-glycyl-*d*-leucine by neither. On Waldschmidt-Leitz and Schlatter's hypothesis the latter should have been hydrolysed by erepsin.

Dakin and Dudley [1913] found that casein and caseose which had been partially racemized by alkali were completely resistant to hydrolysis by pepsin, " trypsin," and erepsin. The racemization had reduced the rotatory power to about 60 per cent. of its original value, and probably only affected the asymmetry of amino-acid residues whose carboxyl groups were linked. On the other hand, Lin, Wu, and Chen [1928] found that partially racemized casein and egg albumin were digested both by trypsin and pepsin, though more slowly and probably less completely than the natural products. This would be expected from the behaviour of polypeptides.

Hydrolysis of Peptides Containing Residues other than α-Amino-acids.

In the first place a number of peptides have been tested in which the amino group of one residue is not in the α-position. As stated above, peptides in which the free amino group is the secondary amino group of proline appear to be hydrolysed by a special enzyme. Waldschmidt-Leitz, Schaffner, and Klein [1928] found that animal erepsin can hydrolyse alanyl- and leucyl-β-amino-butyric acids, while Abderhalden and Reich [1928] found that neither animal "trypsin" (i.e. carboxy-peptidase) nor erepsin attack glycyl-β-alanine and leucyl-β-alanine. Since, however, Abderhalden and Fleischmann [1928] found no hydrolysis by erepsin of leucyl-β-amino-butyric acid, the difference may depend on conditions of working. Trypsin will hydrolyse β-amino-butyryl-*l*-leucyl-tetraglycine [Abderhalden and Fleischmann, 1928], but neither it nor erepsin attack a number of shorter peptides in which the free amino group is in the β-position. Similarly, although Waldschmidt-Leitz, Schaffner, and Klein [1928] found that animal erepsin hydrolyses leucyl-ε-amino-caproic acid, the yeast peptidases do not attack leucyl-γ-amino-butyric acid [Abderhalden and Tateyama 1926], alanyl- and leucyl-δ-amino-valerianic acid [Abderhalden and Hartmann, 1927] or glycyl-ε-amino-caproic acid [Grassmann and Dyckerhoff, 1928, 1]. There is as yet no clear evidence as to the hydrolysability of peptides in which the δ-amino group of ornithine or the ε-amino group of lysine is combined.

Besides non-α-amino acids a large number of other molecular types may form hydrolysable peptides. These are considered in the next section.

Ranges of Specificity of the Different Peptidases.

In what follows it must be remembered that animal erepsin was thought up to 1929 to be a single enzyme, whilst it is now believed to be a group of three or four. Good data exist as to the ranges of gut-erepsin (as a group), trypsin+carboxy peptidase (activated and without kinase), yeast dipeptidase, and yeast polypeptidase. One simple generalization can at once be made. No enzyme has been found which will hydrolyse a peptide not possessing a free amino *or* carboxyl group. Examples of unhydrolysed substances are carbethoxy-tetra-glycyl-amide, and diketopiperazines such as 2, 5-dioxo-piperazine or glycine anhydride, $HN\langle\begin{matrix} CH_2-CO \\ CO-CH_2 \end{matrix}\rangle NH$, and 2, 5-dioxo-3-methyl-6-benzyl-piperazine or alanine-phenylalanine anhydride. Before, however, it is assumed that such rings do not occur in normal proteins, it is desirable that the digestibility of substituted diketopiperazines with free amino or carboxyl groups should be tested.

Yeast dipeptidase attacks 7 dipeptides [Grassmann and Dyckerhoff, 1928]. It hydrolyses none of their derivatives (peptamines, amides, acylated dipeptides, and tripeptides) nor simple amides such as leucine-amide. Hence both free NH_2 and COOH groups are necessary. As, however, *l*-leucyl-*d*-glutamic acid is rapidly hydrolysed in nearly neutral solutions where almost all its molecules have a net charge of one electron, this does not mean that only uncharged molecules are attacked, but probably that the substrate must be of formula

$$\overset{+}{H_3N}-\underset{H}{\overset{R_1}{C}}-CO-NH-\underset{H}{\overset{R_2}{C}}-\overset{-}{COO},$$

the charge of R_2 being irrelevant.

Yeast polypeptidase does not hydrolyse dipeptides, but attacks 7 tripeptides investigated and 5 higher polypeptides, the largest being leucyl-hepta-glycine [Grassmann and Dyckerhoff, 1928]. It digests the esters, not only of tripeptides, but even of dipeptides; amides, such as glycyl-leucine-amide, and even glycine and leucine amides; and peptamines such as glycyl-decarboxy-leucine. On the other hand, it has no effect on acylated peptides such as benzoyl-glycyl-glycine (though Utzino [1928] found that a yeast extract hydrolysed phthalyl-diglycine and phthalyl-triglycine). To sum up, it will attack the peptide linkage provided the substrate molecule possesses a free amino group, and does not possess a free carboxyl in the immediate neighbourhood of the peptide linkage in question.

Animal erepsin and erepsin-free trypsin have now been tested on a very large number of peptides and their derivatives, notably by Waldschmidt-Leitz, Klein, and Schaffner [1927, 1928], Waldschmidt-Leitz, Schaffner, Schlatter, and Klein [1928], Waldschmidt-Leitz and Klein [1928] and Abderhalden [1928, and 1929] with Köppel, Brockmann, Sickel, Rossner, Schwab, Fleischmann, Reich, and many others.[1] Other valuable data on erepsin are given by Levene,

[1] In *Fermentforschung*, vols. 9, 10 and 11.

Steiger, and Bass (1929), Euler and Josephson [1926], and others.

Animal erepsin behaves like a mixture of the yeast peptidases. For its action a free amino group is necessary, a free carboxyl group is not. It attacks the large majority of dipeptides so far investigated at pH 8. The exceptions are aspartyl-glycine, glycyl-aspartic acid, leucyl-glutamic acid, and leucyl-oxyproline. The first three are electronegative, and one is known (see p. 23) to be attacked at a more acid pH. It also attacks amides, e.g. leucine-amide, peptamines such as alanyl-decarboxy-leucine and leucyl-tyramine, and esters such as the "biuret base," triglycyl-glycine ester. More remarkably, it hydrolyses such compounds as alanyl-colamine, alanyl-aniline, glycyl- and diglycyl-*p*-aminobenzoic acids, and even glycyl-, diglycyl-, and triglycyl-arsanilic acids.

Polypeptides composed of leucine and glycine residues are hydrolysed, up to the octodecapeptide leucyl-triglycyl-leucyl-triglycyl-leucyl-nonoglycine (molecular weight 1013). But polypeptides containing certain other residues begin to pass out of its range, at least at pH 8, with increasing molecular weight. Thus dialanyl-cystine and (though more slowly) even di- (alanyl-valyl-alanyl)-cystine are hydrolysed, but di- (leucyl-alanyl-valyl-alanyl)-cystine is not. Tyrosine and other relatively electronegative residues hinder its action (at pH 8). Thus, though erepsin hydrolyses leucyl-glycyl-tyrosine, it does not attack leucyl-triglycyl-tyrosine.

If the free amino group is substituted, as in acetyl-glycyl-glycine, chloracetyl-glycyl-glycine, or N-methyl-leucyl-glycine, erepsin has no action, though so strongly electronegative a substance as glycyl-3, 5-dichlorotyrosine is still hydrolysed.

The problem of the specificity of the peptidase associated with trypsin is complicated by the fact, discovered by Abderhalden and Zeisset [1929], that its preparations contain an enzyme, perhaps identical with histozym, which hydrolyses such substances as chloracetyl-valine and α-bromisovaleryl-di-glycyl-glycine. It is probably not identical with histozym (hippuricase), as Gulewitch [1899] found no histozym in fairly crude pancreatic extracts containing trypsin. This enzyme can be obtained from trypsin preparations, and has no action on proteins or peptides. It is probably the same enzyme which acts on substituted peptides such as carbethoxy-leucyl-glycine and phenyl-carbamide-glycyl-leucine [Abderhalden and Schwab, 1929]. Trypsin preparations vary in the amount of hydrolysis which they produce in such substances, but none have yet been obtained incapable of attacking them. Waldschmidt-Leitz and his colleagues found that kinase did not increase the activity of trypsinogen preparations on carbethoxyl-glycyl-leucine, and on acetyl-phenylalanyl-alanine, but did so on β-naphthalene-sulphonyl-tyrosine, and was necessary for any action at all on benzoyl-glycyl-glycine. It would seem that all the trypsin preparations tested on these acylated peptides were prepared from whole pancreas, as opposed to pancreatic juice.

Activated " trypsin " (i.e. carboxypeptidase) affects a number of simple peptides. Dipeptides are rarely attacked, exceptions being glutamyl-tyrosine and phenylalanyl-arginine. It also attacks a much larger proportion of polypeptides. Tripeptides containing tyrosine, e.g. glycyl-alanyl-tyrosine and glycyl-tyrosyl-glycine are partially hydrolysed, as are some others, e.g. leucyl-glycyl-leucine. On the whole hydrolysability by " trypsin " increases with size. Thus di-alanyl-cystine and di-leucyl-cystine are not attacked by " trypsin," di-(leucyl-glycyl)-cystine and similar pentapeptides, and higher derivatives, such as di-(leucyl-alanyl-valyl-alanyl)-cystine are also vigorously hydrolysed. On the other hand, the largest known polypeptide molecule, leucyl-triglycyl-leucyl-triglycyl-leucyl-nonoglycine, is not attacked. Kinase is sometimes, but by no means always, required for the hydrolysis of these substances.

Hydrolysis begins at the end of the molecule which possesses a free carboxyl. Thus " trypsin " rapidly liberates tryptophan from *l*-leucyl-pentaglycyl-*l*-tryptophan, while erepsin does not do so within the same period, although about equally active in liberating amino nitrogen. A free carboxyl group appears to be needed for " tryptic " action. Thus amides of certain of its substrates are not attacked. As stated above, it seems probable that a free amino group is not necessary.

On the whole the results agree with Northrop's view (see p. 22) that the union between " trypsin " and its substrate is salt-like, the substrate being negatively charged. But this will not explain the hydrolysis of phenyl-alanyl-arginine, though, of course, the guanidine group is at some distance from the peptide linkage attacked. It is, however, probable that other factors than charge influence tryptic digestibility.

Proteinases do not in general attack synthetic peptides. However, Willstätter and Grassmann [1924] found that papain in presence of HCN (though not alone) will hydrolyse leucyl-glycyl-glycine, though not glycyl-glycine or leucyl-glycine. And Abderhalden and Schwab

[1930] found that pepsin hydrolyses di-leucyl-thyroxine. As one of the leucyl residues is attached as an ester this may possibly be due to lipase. Some of the peptidases of animal tissues are probably identical with components of erepsin.

Ranges of Specificity of the Proteinases.

It will be seen from Table XVI. that the ranges of the proteinases are limited, and that their activators or kinases may increase them both upwards and downwards. It may be added that attempts to fractionate them, e.g. to separate a casein-splitting and a gelatin-splitting component of trypsin, have so far failed. There is so far no conclusive evidence that any of them will attack any linkage other than the peptide linkage, although it must be borne in mind that groups (presumably -SH) reacting with nitroprusside are unmasked during peptic digestion [Harris, 1923].

Specificity of the Amylases.

The following substrates are hydrolysed by amylases :—

Amylose and the closely related isolichenin.

Amylopectin and the closely related glycogen.

Dihexosan, trihexosan, tetrahexosan, hexahexosan.

Amylobiose and amylotriose.

Dextrins of imperfectly known nature.

Since emulsin, pancreatic juice, and malt extract all hydrolyse salep mannan [Pringsheim and Genin, 1924], it is just conceivable that this action may be due to amylase. Further a nitrogenous tri-saccharide obtained by Fränkel and Jellinek [1927] from the albumins of egg white and yolk is hydrolysed by saliva (presumably by the amylase). This substance is non-reducing, and yields glucosamine and mannose on hydrolysis.[1] Its bearing on the amylase problem is not clear. Methylated starches and hexosans are not hydrolysed [Kuhn and Ziese, 1926].

It has been shown that lichenin and the polyamyloses when they are broken down by amylase preparations, are attacked by an enzyme other than amylase [Leibowitz and Mechlinski, 1926]. The starch liquefying component of amylases appears to act by breaking down the electrolyte amylopectin [Samec, 1927], and seems not to be identical with the sugar-producing amylase, from which it differs in its heat stability, and otherwise.[2] So far three or four distinct types of

[1] See Levene and Mori [1929].

[2] Ohlsson (1930) has nearly completely separated the two in malt extract. The latter may be amylase + complement. (See p. 140.)

hydrolysis of starch at room temperatures are known; there are also differences in the amount and nature of the dextrin left at the end:—

(*a*) Breakdown to α-maltose by pancreatic and *Aspergillus* amylase (takadiastase). [Kuhn, 1925, 1.]

(*b*) Breakdown to β-maltose by malt amylase [Brown and Morris, 1885]. [Kuhn, 1925, 1.]

(*c*). Breakdown to glucose and a little trisaccharide by "biolase," a maltase -free commercial amylase probably of bacterial origin [Pringsheim and Schapiro, 1926], and by a maltase-free amylase of *Saccharomyces Ludwigii* [Gottschalk, 1926].

(*d*) Breakdown to glucose and a sugar thought to be amylotriose (in the case of glycogen) [Lohmann, 1926].

From the figures given by Lohmann for the specific rotations of his trisaccharide and its hendeca-acetyl derivative, it may quite possibly be found to be identical with that formed by biolase, rather than with amylotriose. If so, muscle amylase falls into Class (*c*). (*b*) and (*c*) are modifiable by high temperature. At 70° malt diastase (replaced at half-hourly intervals to overcome inactivation) gives a high yield of a disaccharide from dephosphorylated amylopectin (Ling and Nanji's α-β-hexaamylose). At 55° maltose and another disaccharide "isomaltose" are formed from amylopectin [Ling and Nanji, 1923, 1925.] Similarly "biolase," which is fairly thermostable, yields a trisaccharide with a little glucose at 70° [Pringsheim and Schapiro, 1926]. The two trisaccharides are probably identical. The English authors regard this sugar as a β-glucosido-maltose of normal (1, 5) ring structure, since it is converted into glucose and maltose, and its osazone into glucose and maltosazone, by emulsin; while yeast maltase gives an isomaltose and glucose or glucosazone. The Germans assign a *h*-glucose, i.e. (1, 4) glucose residue to it for reasons which will appear later, though it is not amylotriose. The nature of the disaccharide or isomaltose is still uncertain, as it yields the same osazone and other derivatives as maltose [Haworth, 1927]. (Of course cellobiose would be formed from amylopectin preparations containing appreciable quantities of hemicellulose, provided the amylase preparation contained lichenase.)

Types (*a*) and (*b*) of hydrolysis have been distinguished by following the optical rotatory power of starch or glycogen solutions during the process. (Figs. 28, 29.) During hydrolysis by pancreatic amylase the rotation remains for several hours (at 20°) higher than that calculated from the reducing power. On adding soda the values become nearly or quite identical. Except for the maximum and minimum values observed during the early stage the results are what would be expected if the starch is converted into α-maltose, which then mutarotates to α-β-

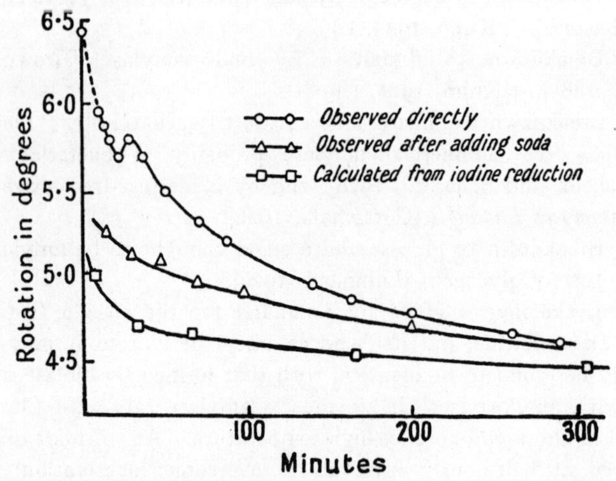

Fig. 28.—Course of hydrolysis of soluble starch by pancreatic amylase. The soda caused mutarotation of α-maltose. [Kuhn, 1925, 1.]

maltose. The optical rotation of α-maltose calculated on this hypothesis agrees within 1·5 per cent. with that calculated by Hudson. The initial disturbance is presumably due to the formation of an

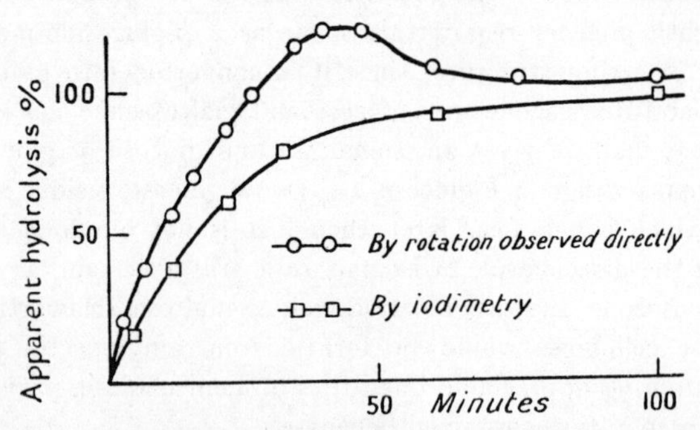

Fig. 29.—Course of hydrolysis of soluble starch by malt amylase. The apparently excessive values are due to β-maltose. [Kuhn, 1925, 1.]

intermediate compound of lower rotatory power than α-maltose. Since the speed with which the minimum is reached varies with the enzyme concentration, the intermediate substance is converted into α-maltose by catalysis, and not mutarotation. Takadiastase behaved like pancreatic amylase save that the initial anomaly was not observed. Malt amylase gave a solution whose rotatory power was raised by alkali, and which agreed quantitatively with the view that β-maltose was first formed and then mutarotated.

Takadiastase was twice as strongly inhibited by αβ-maltose as by β-maltose, as was to be expected if it is inhibited by α-maltose which it forms, but not by β-maltose. α-glucose and β-glucose both inhibited it about equally. Pancreatic amylase was not investigated in solutions more acid than pH 6·0, so that mutarotation of the added sugars was fairly rapid. No definite difference was observed between the inhibitory effect of α and β forms of sugars.

Malt amylase was generally inhibited by β-maltose and β-glucose more than by the α-sugars, but with one specimen the effects were nearly equal, and in another indistinguishable.

All the amylases so far investigated attack dihexosan and trihexosan (at least in the presence of complement (see Chap. VII.)), though emulsin amylase only degrades trihexosan to glucose and a dihexosan according to Pictet and Salzmann [1924]. Amylobiose, which is probably 5, β- (1, 5) glucosido- (1, 4) glucose, is attacked by malt extract but not by pancreatic or salivary amylase or the weak amylase (probably an α-amylase and not the same as β-glucosidase) of emulsin [Pringsheim and Leibowitz, 1925, 1, 2], nor by biolase [Pringsheim and Schapiro, 1926], while muscle amylase does not attack amylotriose [Lohmann, 1926]. On the other hand, maltase-free pancreatic amylase preparations and emulsin, as well as takadiastase and maltase-containing malt extract, hydrolyse the trisaccharide formed by biolase. As it is attacked by maltase-free α-amylase it may be α (1, 4) glucosido-maltose.

When starch or its components are attacked by any of the following mixtures (all maltase-free), glucose in greater or less quantity is produced[1] at a suitable pH :—

Malt amylase + pancreatic amylase.
,, ,, + emulsin.
Pancreatic amylase + salivary amylase.

[Pringsheim and Leibowitz, 1925, 1, 2.] Malt and salivary amylases did not have this effect in combination.

The original theory of the authors was that a combination of α-amylase and β-amylase should yield glucose. But salivary amylase, while agreeing with α-amylase in its failure to hydrolyse amylobiose,

[1] Rona and Hefter [1930] were quite unable to confirm these observations.

agrees with β-amylase (malt extract) in producing glucose in combination with pancreatic amylase. It has not yet been polarimetrically investigated, but in its behaviour towards pH, anions, and poisons, it is identical with pancreatic amylase.

It is rendered extremely probable by the production of maltose in quantitative yield from amylotriose and trihexosan, and also by the formation of either α- or β-maltose, that starch and glycogen are not built from maltose residues, and that the formation of maltose by amylase is at least in part a synthetic process. The synthesis is, however, an irreversible process quite independent of substrate concentration. In some amylase preparations this synthesis is disturbed, and glucose or a trisaccharide may be formed.

Assuming the structural formulae given in the Appendix, the substrates of amylase only agree in possessing glucosidic glucose residues with a (1, 4) oxygen ring. The fourth C atom probably forms one end of a bridge in dihexosan and trihexosan, and may do so in the starch components, but cannot in the glucosidic residue of amylobiose. Hence it is plausible that amylases are specific for (1, 4) glucosides. If so, one would expect amylases to attack methyl (1, 4) glucoside (γ methyl-glucoside). They do not do so, though yeast and liver extracts have a slight action [Kuhn and Wagner-Jauregg, 1926]. But amylase may be more specific (like malt as compared to yeast maltase), and be restricted in its action to (1, 4) glucosides of glucose or its derivatives. Amylase attacks with approximately equal speeds colloidal and crystalloidal substrates, although some at least of the latter are only attacked in presence of complement. Such a property is hard to explain on the view that the enzyme-substrate union is an adsorption. Amylase may adsorb a layer of dihexosan molecules, but can hardly adsorb a layer of starch molecules, which are very large.

Specificity of Zymase.

Yeast ferments the following simple sugars :—

d-glucose, *d*-mannose, *d*-fructose, *d*-mannononose, glyceric aldehyde, and dihydroxyacetone.

The fermentation of the latter produces (2, 5) fructose diphosphoric acid [Lebedev and Griaznov, 1912], and therefore synthesis of a hexose is a preliminary to fermentation. That of glyceric aldehyde, which is very slow, does not apparently involve hexose-phosphate synthesis. It is, therefore, likely that the trioses require for their fermentation catalysts other than those acting on the hexoses, and the same may be true of mannononose. Galactose is fermented by certain yeasts only. No other simple sugars are attacked, although the following have been tested :—

d-erythrose, *d*-erythrulose.
l-arabinose, *d*-arabinose, *l*-xylose, *d*-lyxose.
d-rhamnose.
d-gulose, *d*-talose, *d*-sorbose,[1] *d*-*l*-gulose, *d*-tagatose.[2]
l-glucose, *l*-mannose, *l*-gulose, *l*-galactose, *l*-fructose.
α-glucoheptose, *d*-mannoheptose, *l*-mannoheptose, perseulose, sedoheptose, *d*-mannoketoheptose.
α-glucooctose, *d*-mannooctose.
α-glucononose, α-glucodecose.
[Fischer and Thierfelder, 1894, *Biochem. Handlexikon, passim.*]

d-glucose, *d*-mannose and *d*-fructose all possess the structure

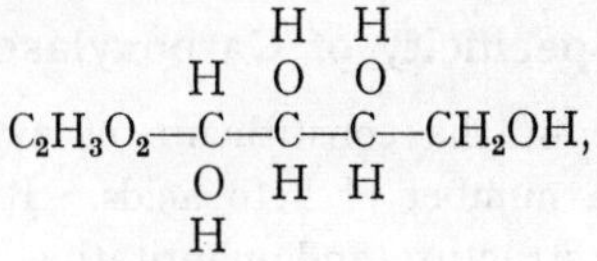

$$\begin{array}{ccccccccc} & & & & H & & H & & \\ & & H & & O & & O & & \\ C_2H_3O_2 & — & C & — & C & — & C & — & CH_2OH, \\ & & O & & H & & H & & \\ & & H & & & & & & \end{array}$$

and have a common hypothetical enolic form which may be an intermediate stage in their fermentation; *d*-galactose has the structure

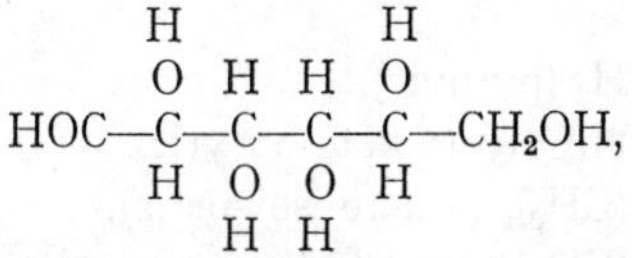

$$\begin{array}{ccccccccccc} & & H & & & & & & H & & \\ & & O & & H & & H & & O & & \\ HOC & — & C & — & C & — & C & — & C & — & CH_2OH, \\ & & H & & O & & O & & H & & \\ & & & & H & & H & & & & \end{array}$$

i.e. differs from the above sugars with regard to its fourth C atom. It is remarkable that *d*-talose and *d*-tagatose, which are related to it as are *d*-mannose and *d*-fructose to *d*-glucose, should be unfermentable.

None of the simple derivatives of the fermentable sugars, except α-glucosides, phosphoric esters, and disaccharides are fermentable. It was generally supposed that hydrolysis was a preliminary to the fermentation of maltose, sucrose, and lactose, but Willstätter and Steibelt [1921] find that some yeasts which are poor in maltase ferment that sugar far more rapidly than their extracts can hydrolyse it, while Willstätter and Oppenheimer [1922] observed lactose fermenta-

[1] Fermented by an Italian yeast (*Mezzadroli Chem. Zentbt.*, 1919, II. 500).
[2] Cf. Lindner (*Chem. Zentbt.*, I., 56).

tion by lactase-free yeasts. Willstätter and Lowry [1925] extended the principle to sucrose. It is, of course, conceivable that the failure to obtain maltase and lactase from these yeasts may have been due to special circumstances, and that it may have been present in the living cells.

Willstätter and Sobotka [1922] found that although yeast fermented α- and β-glucose with equal rapidity, yet from an equilibrium mixture of the two, α-glucose was preferentially fermented, just as glucose is preferentially selected from a mixture of glucose and fructose though both ferment at the same rate. Such phenomena as these (of which other examples are given in Harden's monograph in this series) must remain obscure until the individual constituent enzymes of zymase have been isolated.

The specificity of muscle " zymase " depends mainly on the presence of coenzymes and activators, and will be considered in Chapter VII.

Specificity of Carboxylase.

Carboxylase (one of the constituents of zymase) catalyses the decarboxylation of a number of keto acids. Its field of activity is limited absolutely by structure, and quantitatively by stereochemistry. Writing R for the group $>$CH—CO—COOH, the following acids are decarboxylated :—

RH_2 (pyruvic).
$RH(CH_3)$ (α-keto-butyric).
$R(CH_3)_2$ (α-keto-isovaleric).
$R(CH_3)(C_2H_5)$ (α-keto-capronic).
RH(COOH) (oxalacetic).
$RH(CH_2COOH)$ (α-ketoglutaric).

[Neuberg and Weinmann, 1928.] Both components of *dl*-methyl-ethyl-pyruvic (α-keto-capronic) acid are decarboxylated, the dextrorotatory more rapidly [Neuberg and Peterson, 1914]. In the case of oxalacetic acid, not only the salts, which are regarded as keto-compounds, but the two enolic forms of the free acid, namely oxymaleic and oxy-fumaric, are rapidly attacked [Mayer, 1913]. On the other hand, trimethyl-pyruvic acid is quite resistant [Neuberg and Weinmann, 1928]. It is suggested that this may be due to its inability to form an enolic isomer, the enolic forms being the actual reactants.

Specificity of Xanthine Oxidase.

This question has been investigated by Dixon [1926] and Coombs [1927]. The enzyme catalyses the donation of two or four H atoms from purine derivatives and aldehydes to a variety of acceptors. With regard to acceptors the enzyme exhibits no specificity. The following are reduced :—

O_2, H_2O_2, $—NO_3'$, I_2, $—MnO_4'$, dinitrobenzene, picric acid, quinone, alloxan, guaiacum blue, thionine, methylene blue, and all Clark's reduction potential indicators.

Hence the *r*H developed (which cannot be directly measured with an electrode) is less than 7. The only hydrogen acceptor tested which was not reduced was the oxidized form of glutathione, which requires a very powerful reducing agent (low *r*H). A comparison of the rates of reduction of the Clark indicators by hypoxanthine showed that this depended mainly on the number of sulphonic groups in the dye molecule (and hence its charge) rather than on the *r*H at which it is reduced, which determines the free energy of the action. Sulphonation had an inhibitory effect.

The possible hydrogen donators may be grouped into five classes :—

(*a*) Hydrogen donators rapidly oxidized :—

Xanthine, hypoxanthine, 6, 8-dihydroxypurine, 2-thioxanthine.

(*b*) Hydrogen donators slowly oxidized [1] :—

Acetaldehyde, formaldehyde, benzaldehyde,[2] salicylaldehyde [2] piperonal,[2] adenine.

The reduction of adenine is possibly due to impurities or to the presence of adenase. There is no doubt of its inhibitory action, so it may fall into class (*c*).

(*c*) Substances which do not donate hydrogen, but inhibit strongly (see Chap. III.) :—

Guanine, uric acid, 3-methylxanthine, 1-methylxanthine, 1-methylguanine, 7-methylguanine, 1, 7-dimethylguanine.

(*d*) Substances which do not donate hydrogen, and inhibit moderately :—

8-methylxanthine, 9-methylxanthine, alloxan (but see later comment).

[1] Wieland and Rosenfeld (1929) state that they have partially separated xanthine oxidase and aldehyde oxidase. The question of their unity, like that of the dipeptidases, seems to be still open.

[2] Personal communication of Dr. Dixon.

(*e*) Substances which do not donate hydrogen, and inhibit very slightly or not at all :—

1, 3-dimethylxanthine, 3, 8-dimethylxanthine, theobromine, (3, 7-dimethylxanthine), caffeine (1, 3, 7-trimethylxanthine), thymine, cytosine, uracil, histidine, benzimidazole.

(*f*) The following substances, among others, do not donate hydrogen, but it is not known whether they inhibit the action, though there is no reason to suspect it :—

Acetone, formate, acetate, lactate, pyruvate, succinate, oxalacetate, citrate, glucose, lactose, lower amino-acids, including tryptophan, pure peptone, benzylamine, quinoline, morphine, ricin.

Taking the rate of reduction of methylene blue by xanthine as 100, the following are the rates for the other substances, all in ·009 per cent. solution :—

Hypoxanthine	200
6-8-dihydroxypurine . . .	90
2-thioxanthine	64
Adenine	2
Acetaldehyde	0·16 (calculated).

As one molecule of hypoxanthine reduces two of methylene blue, the first figure should be halved for comparison.

It will be seen that aldehydes are slowly oxidized, but rapid oxidation is only found with the oxypurines, a thiopurine, and possibly adenine (6-aminopurine). 2, 6, and 6, 8-dihydroxypurines are oxidized at about equal rates, and the substitution of a S atom for one of the O atoms in xanthine has little effect. Union with the enzyme, as shown by inhibitory action, is furthered by amino groups, and discouraged by methyl groups. These influences can roughly neutralize one another, as in the case of the methyl-guanines. Dimethyl compounds have less inhibitory action (i.e. affinity for the enzyme) than monomethyl compounds. Although an amino group increases affinity it prevents oxidation, as is shown by comparison of hypoxanthine and its amino derivative guanine. The whole purine nucleus is required for union, except perhaps in the case of alloxan. The inhibition by this substance may, however, be explained by the fact that it can act as a hydrogen acceptor, and therefore competes with the methylene blue for hydrogen, not with the purines for enzyme. Other pyrimidine and iminazole derivatives have no affinity for the enzyme.

Specificity of Bacterial Dehydrogenases.

These catalysts have been studied very thoroughly by Quastel and his colleagues [1925-28]. Resting, i.e. non-multiplying *B. coli* reduces a very large number of substances, and the problem of specificity is not yet solved. A dehydrogenase specific for lactic acid and a few similar substances can be obtained in solution [Quastel and Wooldridge, 1928], but the other enzymes have not yet been separated.

One method employed in investigating their specificity was as follows. Two substances A and B are added in appropriate concentrations, and the times t_A and t_B taken by them to reduce a certain amount of methylene blue noted. The time t_{A+B} taken by both together is then compared with these. If V_A, V_B, V_s be the reciprocals of these times, then (if t_A and t_B are both finite), if the two substances are activated by the same enzyme, V_s must lie between V_A and V_B; if they are quite independent $V_s = V_A + V_B$. In other cases V_s lies between $V_A + V_B$ and the greater of V_A and V_B, and it is possible to calculate approximately how far a given substrate interferes with the oxidation of any other.

Again the effect of different inactivators on different oxidations may be compared. These inactivators may either be substances analogous to the substrates (e.g. oxalic acid) or "poisons," e.g. toluene, $KMnO_4$ or $NaNO_2$.

The toluene-treated organism possesses groups which activate succinic acid, α-hydroxy-acids, and formic acid. These three are nearly if not quite independent, as judged by the above tests. The normal organism can activate a great variety of other substances as well. Among these it is quite clear that glucose, mannose, fructose, galactose, mannitol and sorbitol are all activated by the same catalyst. But the sharp distinction between different catalysts which is characteristic of the toluene-treated organism no longer appears. Thus the combined velocity of reduction of methylene blue by lactic and succinic acids together is much less than the sum of the two alone. Again the inhibitory effect of oxalic acid on lactic acid oxidation is relatively small (see Chap. III.). Either in the normal organism the catalysts found on the toluene-treated organism have a far wider range of specificity, or the former possesses a number of additional catalysts of low specificity. Quastel and Wooldridge [1927, 2] incline to the former view as the result of a study of inactivation by a variety of different agents.

Specificity of Succinodehydrogenase.

This enzyme has a high specificity as regards reducing substrates. Thunberg's [1917] mammalian muscle preparations had no action on the vast majority of possible hydrogen donators, including mono-carboxylic acids, and such dicarboxylic acids as malonic, fumaric, and glutaric. A slow action was, however, observed with methyl-succinic (pyrotartaric) acid, ethyl succinate, and ethyl aceto-succinate.

While the ordinary preparations reduce O_2 as well as methylene blue, this is probably due to the presence of an oxygen activator (see Chap. VII.). Further details with regard to the specificity of bacterial succinodehydrogenase are given on page 62. It is not identical with fumarase, being more thermolabile [Alwall, 1928], and the work of Woolf [1929] shows that it is different from aspartase.

Specificities of Other Dehydrogenases.

Mammalian tissues, bacteria, and yeast, catalyse a large number of oxidations by methylene blue. However, only a few of the de-hydrogenases concerned have been separated from one another and obtained in solution. The fullest study of specificity is that of Bernheim [1928, 1, 2] on potato aldehyde oxidase, yeast lactic dehydro-genase, and liver citric dehydrogenase. Solutions of these can be obtained which do not activate the majority of H donators activated by muscle.

Yeast lactic dehydrogenase [Bernheim, 1928, 2] reduces methylene blue, but not O_2, H_2O_2, even in presence of peroxidase, or nitrates. It is thus far more specific than xanthine oxidase. It oxidizes lactic, and rather more slowly, α-hydroxybutyric acids, but no others of a large number tried. Pyruvic, oxalic, and glyceric acids inhibit. The properties of the lactic dehydrogenase which can be extracted from *Bacillus coli*, so far as they have been studied, are exactly the same [Stephenson, 1928].

Liver citric dehydrogenase [Bernheim, 1928, 2] reduces methylene blue and *m*-dinitrobenzene, but not O_2 or nitrates. It will oxidize no substance so far tried except citric acid. The oxidation is in-hibited by aconitic acid (citric acid $- H_2O$) but not by any of the other substances tried, e.g. succinic, tartaric, or lactic acids.

Potato aldehyde oxidase [Bernheim, 1928, 1] reduces methylene blue, Clark's indicators, quinone, *m*-dinitrobenzene, and nitrates. It probably reduces O_2, but it is so rapidly destroyed by O_2 + aldehyde, even in presence of catalase, that this is uncertain. It oxidizes nine aldehydes, but not alcohols, ketones, sugars, acids, including formic and pyruvic, nor a variety of other reducing agents, including hypoxanthine.

Specificity of some other Oxidases.

Oxygenase has an affinity for CO of the same order of magnitude as that for O_2 [Warburg, 1927], though less. Peroxidase will act on a very large variety of reducing substrates, including iodides as well as organic compounds. Besides H_2O_2 it activates persulphate [Dixon, 1929], ethyl hydrogen peroxide and peracetic acid, but is inhibited by dioxymethyl peroxide and disuccinyl peroxide, while diethyl peroxide does not interact with it [Wieland and Sutter, 1930].

Stereochemical Specificity as Shown by Reaction Products.

A group of enzymes has been studied which produce optically active products, not from racemic mixtures, but from symmetrical molecules. Perhaps the distinction is not ultimate, for so-called symmetrical molecules may include equal numbers of optically active tautomers. Some of these enzymes, for example fumarase, catalyse reversible reactions. If so, they show stereochemical specificity of the ordinary type in catalysing the reverse reaction.

An example is glyoxalase [Dakin and Dudley, 1913, 2, 1914]. This converts glyoxal and substituted glyoxals into the corresponding glycollic acids, R—CO—COH + H_2O = R—CHOH—COOH. Except where R is H, the acids are optically active. In the case of benzyl and isobutyl glyoxals *d*-phenyl-lactic and *d*-leucic acids are formed almost, if not quite, exclusively. But while phenyl and methyl glyoxals both give an excess of lævo-rotatory acids (mandelic and lactic respectively) appreciable amounts of the *d*-acids are formed as well. The stereochemical specificity is thus nearly or quite absolute in the first two cases, quantitative in the second two. This case is paralleled by the specificity of prunase with regard to tertiary alcohol glucosides.

Neuberg [1913] found that yeast-juice gave a mixture of *l*- and *dl*-lactic acids, but live yeast, and various resting bacteria [Neuberg and Gorr, 1925] gave only *dl*-acid.

Fumarase and aspartase, which produce *l*-malic and *l*-aspartic acids respectively from fumaric acid, appear to be absolutely specific

[Dakin, 1922; Quastel and Woolf, 1926; Woolf, 1929]. In these cases the action is reversible, and *l*-malic and *l*-aspartic acids are readily attacked by these enzymes. *l*-malic acid is selectively removed from a racemic mixture, but it has not been shown that *d*-malic acid is quite unaffected. Nor has *d*-aspartic acid been tested. If, however, the specificity were not absolute, it is obvious on thermodynamical grounds that the enzyme would finally racemize malic acid.

The most extensively investigated case of this kind is that of the formation of mandelonitrile from HCN and benzaldehyde under the influence of the oxynitrilase of emulsin. This has been investigated by Rosenthaler [1908, 1909, 1913, 1922], Bayliss [1913], Krieble [1913], Krieble and Wieland [1921], and Nordefeldt [1921, 1922, 1923].

The history of this synthesis bristles with controversy. The complications are mainly due to the following facts. *dl*-mandelonitrile is formed even in the absence of enzyme, at a rate depending on pH, and nearly proportional to the OH concentration. Thus Nordefeldt found that the rate was increased 33,000 times on increasing the pH from 2·9 to 8·0, the theoretical ratio being 126,000. Hence, in an unbuffered system, the enzyme preparation will affect this catalysis as well as the specific catalysis. As the equilibrium constant varies only slowly with pH or temperature the velocity of the reverse reaction also varies greatly with pH. Again benzoic acid is formed by autoxidation of benzaldehyde, which further upsets the course of the reaction. *d*-mandelonitrile racemizes spontaneously. The speed of racemization has a Q_{10} of about 3·2, and increases with pH, becoming almost instantaneous at pH 7. Finally, almond emulsin catalyses the formation of the *d*-nitrile, peach and cherry leaf extracts that of the *l*-nitrile. Rosenthaler believed that he had separated a synthetic and hydrolytic component of emulsin, but later workers were unable to substantiate his claim, and it appears that the final equilibrium reached is independent of the presence or amount of emulsin in the system, provided this does not affect the pH, and hence the amount of HCN, as opposed to CN′ ions in the system.

It is clear, then, that in order to study the activity of the enzyme itself, relatively undisturbed by non-enzymatic reactions, the reaction must be studied at a low temperature and pH. Room temperature and pH 4·5 are suitable, though the optimum pH for the enzyme is 6 or greater. When the pH or temperature is raised the optical activity of the system rises rapidly to a maximum, and then declines owing to the production of *dl*-mandelonitrile. When the reaction was carried out at 0° and stopped before equilibrium was quite reached, so as to avoid racemization, Krieble and Wieland were able to obtain 100 per cent. pure *l*-mandelonitrile with an enzyme from peach-leaves. This enzyme at least appears to be absolutely specific. There is no reason to doubt the absolute specificity of other oxynitrilases, but it has not been so conclusively proved.

Single and Double Specificity.

The large majority of reactions catalysed by enzymes can be expressed as $A + B \rightleftharpoons C + D$, the back reaction being sometimes hypothetical. In the case of hydrolytic enzymes B is probably always H_2O, though it is possible that emulsin may catalyse such reactions as saligenin β-glucoside + methanol ⇌ saligenin + β-methylglucoside without the intervention of water, and similar possibilities exist for other reactions. It is an open question whether the special position of water is due to its affinity for the hydrolytic enzymes, which activate it, or whether it merely bombards activated substrate molecules (see p. 182). The nature of the specificity for A is best seen by a consideration of C and D. In some cases either C or D may be one of a number of different substances. Thus the lipases hydrolyse the esters of a vast variety of alcohols and acids. In others C is a single substance or one of a small group, but D has a wide range. Prunase, for example, will only hydrolyse β-glucosides and β-*d*-epirhamnosides, but shows little specificity for the alcoholic group. Finally both C and D may be single substances or very small groups. Arginase and taka-maltase are examples.

The same classification may be applied to non-hydrolytic enzymes. Carboxylase catalyses a reaction $A \rightarrow C + D$ where D is always CO_2, but C covers a whole class of acids. In the case of dehydrogenases there is usually a small range only of reducing substrates, and a larger one of hydrogen acceptors, but while xanthine oxidase is very unspecific for the latter, lactic dehydrogenase has so far only been found to reduce methylene blue, its natural oxidizing substrate being unknown. It is thus on the way to being an enzyme of double specificity.

The following classification may therefore provisionally be made. But the three classes shade off into one another. A full specification of an enzyme would include values of the Michaelis constants for all substances with which it can react, including H_2O if possible. The nearest approach to this ideal has been made in the case of prunase (see p. 49).

Enzymes of low specificity.

Lipases, proteases, peptidases.

Enzymes of single specificity.

Yeast maltase, yeast saccharase, prunase, almond lactase, kefir lactase, phosphatases, sulphatases, carboxylase, most dehydrogenases, peroxidase.

Enzymes of double specificity.

Malt maltase, taka-maltase, myrosin, arginase, urease, aspartase.

APPENDIX.

Structure of Carbohydrates.

The structural formulæ of certain carbohydrates are still controversial. In many cases the exact structure does not at present matter to enzyme chemists. It is important that the ring structure of fructose is different in α- and γ-methyl fructoside, and the same in the latter as in sucrose, but the question as to the type of ring will only become fundamental when more is known as to enzyme structure. The formulæ here adopted may be regarded as provisional, but reasonably probable.

In Table XVII. Bergmann's notation is used. E.g., 4, α (1, 5) glucosido- β (1, 5) glucose means

```
     ┌─────────────────O──────────────┐
     │                       H        │
     │                       |        │
   H—C──────────┐        HO—C─────────┼────┐
     |          │            |        │    │
   H—C—OH       │         H—C—OH      │    │
     |          │            |        │    │
  HO—C—H        O         HOC—H       │    O, or
     |          │            |        │    │
   H—C—OH       │         H—C─────────┘    │
     |          │            |             │
    HC──────────┘         H—C──────────────┘
     |                       |
     CH2OH                   CH2OH

        CH2OH                            CH2OH
   H    C——O      H                H     C——O       OH
    \  /|    \   /                  |   /|    \    /
     \/ H     \ /                   |  / H     \  /
     C         C(α)                 C           C(β)
     |\  OH  H /|                   |\  OH  H  /|
     | \ |   |/ |                   | \ |   | / |
    OH  C———C   └────────O──────────┘  C———C    H
        |   |                          |   |
        H   OH                         H   OH
```

The number 4 refers to the carbon atom to which the reducing group is attached. In the case of a ring-compound such as dihexosan the number before each sugar refers to the carbon atom to which its reducing group is attached in another molecule. The authority for the majority of the formulæ is that of Haworth and his colleagues [Haworth, 1929], but in some cases further details have been added which are rendered probable from enzyme action. Thus saccharose yields α-glucose on hydrolysis and is hydrolysed by an α-glucosidase. And yeast saccharase hydrolyses the lævo-rotatory component of αβ-methyl (2, 5) fructoside which Schlubach and Rauchalles [1925] regard as the β-component. Hence saccharose is here provisionally called an α-glucosido- β-fructoside, although Haworth regards the configuration of the 1 carbon atoms as an open question. In Haworth's nomenclature (1, 5) and (2, 6) sugars are pyranoses, (1, 4) and (2, 5) sugars furanoses.

TABLE XVII.

Saccharose	1, α (1, 5)	glucosido-1, β	(2, 5)	fructoside.
Trehalose	1, α (1, 5)	glucosido-1, α	(1, 5).	glucoside.
Maltose	4, α (1, 5)	glucosido-	(1, 5)	glucose.
Cellobiose	4, β (1, 5)	glucosido-	(1, 5)	glucose.
Gentiobiose	6, β (1, 5)	glucosido-	(1, 5)	glucose.
Lactose	4, β (1, 5)	galactosido-	(1, 5)	glucose.
Melibiose [1]	6, α (1, 5)	galactosido-	(1, 5)	glucose.
Turanose [2]	6, β (1, 5)	glucosido-	(2, 5)	fructose.
Primeverose [3]	6, β (1, 5)	xylosido-	(1, 5)	glucose.
Amylobiose [4]	5, β (1, 5)	glucosido-	(1, 4)	glucose.
Dihexosan [4]	5, β (1, 4)	glucosido-5, α	(1, 4)	glucoside.
Raffinose	6, α (1, 5)	galactosido-1, α	(1, 5)	glucosido-1, β- (2, 5) fructoside.
Melezitose [2]	6, α (1, 5)	glucosido-	(2, 5)	fructosido-1, α- (1, 5) glucoside.
Amylotriose [4]	5, β (1, 5)	glucosido-5, β	(1, 4)	glucosido- (1, 4) glucose.
Trihexosan [4]	5, β (1, 4)	glucosido-5, β	(1, 4)	glucosido-5 α- (1, 4) glucoside.
Hexose diphosphate [5]	(2, 5)	fructose	(1, 6)	diphosphate.

[1] But cf. Zemplén (1927).

[2] Zemplén (1926, 2). Hudson's rule shows that the first glucose is α.

[3] Helferich and Rauch (1927).

[4] Pringsheim and Leibowitz (1925, 2), (revised in the light of Haworth's work).

[5] Morgan and Robison (1928).

CHAPTER VII.

COENZYMES, ACTIVATORS, KINASES, AND COMPLEMENTS.

THE action of an enzyme is always affected by chemical influences other than substrate concentration and pH. In the case of some enzymes, such as saccharase, any chemical change in the system comprising the fairly pure enzyme at optimum hydrogen ion and substrate concentrations, either leaves the reaction velocity unaltered, or depresses it. In the case of many other enzymes one of two things is found. Either the reaction can be markedly acclerated by the addition of some substance to the enzyme preparation; or the latter can be separated into two or more portions, all of which must be present for full activity (in some cases for any activity at all). Sometimes one of these constituents is found to take an originally unsuspected part in the catalysed reaction, for example, phosphate in alcoholic fermentation. In others this is equally clearly not so, for example, HCN does not intervene as such in papain proteolysis.

It will be convenient to restrict the designation of co-enzyme to heat-stable crystalloidal organic substances of fairly high specificity associated with an enzyme in nature. Thus an active zymase preparation will be said to include :—

1. Several enzymes, all colloidal and heat-labile, probably including one which synthesizes fructose diphosphate and simultaneously divides a hexose molecule in two, an oxidizing-reducing enzyme, carboxylase, and two phosphatases. This list is not necessarily complete.

2. Cozymase, a fairly complex but heat-stable and crystalloidal organic substance.

3. Phosphate.

The word "kinase" was invented to express the view that enterokinase and thrombokinase respectively catalysed the formation of trypsin and thrombin from inactive precursors. This view is probably incorrect. On the other hand, enterokinase being a colloid cannot be classed with coenzymes.

The term "complement" is employed by Pringsheim for substances which extend the action of an enzyme to fresh substrates.

Thus the complement of amylases enables them to attack otherwise indigestible dextrin, that of myozymase [1] (called the activator or hexokinase by Meyerhof) to glycolyse sugars as well as starch and glycogen.

It will be convenient to employ the term activator for non-specific substances which permit or increase the activity of an enzyme. The activation may be primarily physical, as in the case of a variety of colloids which activate lipases, or chemical, as in the case of the anions which activate animal amylases.

But these distinctions are very far from sharp, and only where enzymes have been purified by modern methods are the facts even moderately clear. In many cases we do not know whether an activator is acting directly, or merely competing with the enzyme for inhibitory substances. And since activators commonly alter the pH optimum they may increase enzyme action at one pH, and depress it at another. Much of the literature on enzyme activation is, moreover, of dubious value, owing to the changes of pH produced by the alleged activators.

Phosphatases.

Erdtman [1927] has shown that preparations of kidney phosphatase can be separated into two fractions by dialysis and otherwise. In particular a preparation purified by adsorption with alumina may have its activity increased as much as seven-fold by the addition of dialysate from the crude preparation. Erdtman [1928] showed that the activator consists of Mg̈ ions, which have a notable effect in concentrations as low as 4×10^{-6} M. Cä and Bë in similar concentrations are ineffective, and Zn̈ poisonous. Other activators may exist.

Harden and Young [1906] showed that arsenate activated alcoholic fermentation by dead yeast, in particular the process of hydrolysis of (2, 5) fructose diphosphate, which is the limiting factor in fermentation with zymin or zymase preparations; and Neuberg and Leibowitz [1927, 1928] found that the rate of hydrolysis of fructose diphosphate by killed yeast was about doubled by ·006 M arsenate. The effect on *Aspergillus* phosphatase, though less, was marked. Now, whereas kidney phosphatase acting on (2, 5) fructose diphosphate gives Robison's hexosemonophosphate as an intermediary product, *Aspergillus* taka-phosphatase gives the almost optically inactive Neuberg's monophosphate. Yeast alone gives a mixture of the two, but on addition of arsenate the product consists almost entirely of the Neuberg

[1] This word will be used for the glycolytic enzyme complex of muscle.

ester. The authors therefore conclude that yeast contains two phosphatases, one not affected by arsenates, which yields the Robison ester, and one activated by these salts which yields the Neuberg ester. The work of Meyerhof [1927] makes it fairly certain that the effect of arsenate on muscle fermentation is due to a speeding up of the phosphatase.

Mayer [1928] found that a number of organic arsenic compounds accelerated the hydrolysis of fructose diphosphate by yeast-juice. The results were somewhat variable and the acceleration was always less than in the corresponding amount of arsenate, though in one case sodium phenyl-arsenate was almost as active as the inorganic compound. It is, of course, possible that they were due to the liberation of inorganic arsenic.

Lipases.

The literature with regard to the activation of lipases is enormous. The addition of various substances to pancreatic lipase usually alters the optimum pH. Thus Platt and Dawson [1925] found that the optimum pH for the hydrolysis of ethyl butyrate by pancreatic lipase was 6·9 in phosphate, 8·5 in borate buffer, and that in a borate-phosphate mixture the optimum was the same as in phosphates alone, while Willstätter, Waldschmidt-Leitz, and Memmen [1923] found that calcium oleate, bile salts, and albumin activate in an alkaline medium and depress in an acid, i.e. shift the pH optimum to the alkaline side. On the other hand, crude gastric lipase contains substances which give it a pH optimum in the neighbourhood of 4, whereas, when highly purified, its behaviour with regard to pH is much the same as that of pancreatic lipase [Willstätter and Memmen, 1923]. The activability may depend on the degree of purification. Thus Willstätter and Bamann [1928] found that it was only after a certain degree of purification that either gastric or pancreatic lipase could be activated by bile salts. The activation depends on the substrate. Platt and Dawson compared the effects on a glycerol extract of pancreas, of egg albumin and edestin. They found that albumin enhanced its activity on ethyl butyrate but depressed the hydrolysis of olive oil, whilst edestin enhanced its activity on both substrates. The crude enzyme was more efficient on olive oil, as was the enzyme + edestin, while the albumin rendered it more active on the ethyl butyrate. As the solutions were not buffered, pH effects may have played a part here. Platt and Dawson regard this as the explanation of Falk's [1915] partial separation of *Ricinus* lipase into a fat-splitting " globulin fraction" and an ester-splitting "albumin fraction." Willstätter and his colleagues found that bile salts act like edestin. Trypsin may depress lipase activity by digesting the protein.

Willstätter, Waldschmidt-Leitz, and Memmen [1923] and Willstätter and Memmen [1924, 1] found that calcium salts, bile salts, and albumin in alkaline solutions activate the hydrolysis by pancreatic lipase of olive oil, methyl butyrate, and triacetin; and that of the latter two is also activated by sodium oleate. But proteins inhibit tributyrin hydrolysis, though certain amino-acids and peptides accelerate it [Willstätter and Memmen, 1923]. Curiously enough glycerol, though a product of reaction, accelerates, even in very high concentrations, the velocity of hydrolysis of olive oil reaching a maximum value of about three times that in aqueous emulsion, when the glycerol concentration reaches 70 per cent. An optimum concentration of activator is generally found. Thus Willstätter and Memmen [1923] found that the rate of tributyrin hydrolysis by pancreatic lipase was about doubled by 1·5 mg. per cent. sodium glycocholate, while the velocity fell back towards its original value in higher concentrations.

The action of different activators may or may not be additive. Thus Willstätter, Waldschmidt-Leitz, and Memmen [1923] found the following figures for 13 c.c. of pancreatic lipase solution at pH 8·9-5·5:—

mg. $CaCl_2$.	mg. Na Glycocholate.	mg. Albumin.	Per Cent. Hydrolysis.
0	0	0	4·0
10	0	0	19·4
0	10	0	10·9
10	10	0	13·6
0	0	90	17·4
10	0	15	23·6
0	10	15	20·8

Here calcium and bile salts do not co-operate, but each co-operates with albumin. Willstätter and his colleagues employed $CaCl_2$ and albumin together as their standard activator for lipase estimation. Since it can be shown that either lipase or fat is adsorbed by albumin, Willstätter and his colleagues regard activation as due to the formation of complex adsorbates of enzyme and substrate. It is difficult to see why, on this hypothesis, albumin should inhibit in acid solutions, and hard to explain the action of phosphates and bile salts. It has often been held that the latter act by promoting emulsification of fat, but Rona and Kleinmann [1926] have shown that this has very little influence on the rate of lipase action, and Terroine [1910] has shown that they activate the hydrolysis of soluble esters.

Cozymase.

Harden and Young [1906] showed that zymase (yeast press-juice) can be separated into a colloidal, thermolabile fraction (apozymase), and a thermostable dialysable organic fraction (cozymase). Euler and his colleagues have further investigated the nature of the latter. Euler and Myrbäck [1923, 1] showed that when washed and dried yeast, itself incapable of fermentation, was mixed with varying amounts of cooked yeast-juice either the cozymase or apozymase might be the factor limiting the rate of fermentation. When small quantities of cozymase are added to the dried yeast this rate is at first proportional to the amount of cozymase, and then becomes constant, and the result is the same when small amounts of apozymase are added to cozymase. On the other hand, an amount of cozymase which produces only 30 per cent. activation of a given apozymase is not fully utilized. By adding more apozymase the velocity can be further increased by about 70 per cent.

In this way the unit of cozymase is defined [Euler and Myrbäck, 1923, 2] as the amount needed to produce 1 c.c. of CO_2 per hour from 0·2 gm. washed dried yeast in presence of excess of phosphate and sugar at pH 6·3. While cooked yeast-juices contain 200 to 300 such units per gram, the activity can be greatly increased by purification. Thus Euler and Myrbäck [1928] obtained their most active preparation as follows. The cooked juices are centrifuged, the clear fluid brought to pH 6, and lead acetate added. The filtrate is treated with $Hg(NO_3)_2$, and the precipitate decomposed with H_2S. The resultant solution is carefully treated with a small amount of $AgNO_3$, and then with weak NH_3 solution. The silver precipitate is again decomposed with H_2S. This precipitation may be repeated. The resultant solution is precipitated with phosphotungstic acid. This precipitate is extracted with ether and amyl alcohol in presence of HCl, and the extract evaporated to dryness. The substance so obtained consists in part of adenyl-thiomethylpentose, which is precipitated by picric acid. The resultant preparations contain 50,000 to 83,000 units per gram and only traces of S. They contain 6 per cent. to 7 per cent. P, [Euler and Myrbäck, 1929], yield large amounts of a base, probably adenine, on hydrolysis, and contain 15 per cent. N. On hydrolysis they also give a carbohydrate with pentose reactions. The substance therefore appears to be an adenine-nucleotide.

Other methods of purification were also fairly successful. The cozymase is destroyed by great heat, strong acids and bases, but not by such oxidizing agents as I_2, H_2O_2, or $KMnO_4$, nor by reduction with H_2 and Pt. It is resistant to most enzymes, including trypsin and erepsin, but it is destroyed by lipase preparations [Buchner and Klatte, 1908; Myrbäck and Nilsson, 1927], by an enzyme present in yeast-juice, and by various tissues [Holden, 1924]. However, the only active purified enzyme is kidney nucleosidase [Euler and Myrbäck, 1928]. These authors conclude that cozymase is a nucleoside or nucleotide. By a diffusion-method which is quite independent of the degree of purity of the cozymase, Euler, Myrbäck and Nilsson [1927] found a molecular weight of 486 ± 6. A nucleotide derived from 1 adenine + 2 pentose + $1H_3PO_4 - 3H_2O$ would have a molecular weight of 481.

Euler and Myrbäck [1927] and Myrbäck and Jacobi [1926] have shown that in all probability cozymase is necessary for the aldehyde-mutase of yeast, which catalyses the Cannizzaro transformation of aldehydes into the corresponding esters. Both it and zymase require coenzymes, and the activities of the two coenzymes are not only in the same proportions in yeast and muscle juice, but this proportion is unaltered by various types of preliminary purification. It is, however, conceivable that the latter fact may be due to the identity of co-mutase and adenyl-thio-methyl-pentose. As the pH optima for disappearance of inorganic P and aldehyde mutation are the same, it is also plausible that the same enzyme is concerned in both. Finally, it is highly probable that the same cozymase is required by yeast zymase and myozymase [Meyerhof, 1927]. Their distribution and general properties are similar. Meyerhof [1918] and Euler, Nilsson and Janson [1927] also identify cozymase with the coenzyme of respiration. The ratio of the two substances in yeast-juice and muscle-juice is the same, and both are heat-stable, but the evidence for their identity is incomplete. Cozymase is found in most mammalian organs and blood. It is not identical with insulin or glutathione.

Amylases and Salts.

There is no doubt that, at a pH in the neighbourhood of 7, NaCl greatly increases the activity of a group of amylases of animal origin. However, there is a serious dispute as to whether the enzymes retain any activity at all when free of all salts. Cole [1904], Michaelis and Pechstein [1914], and other authors have found dialysed ptyalin quite inactive on dialysed soluble starch. Myrbäck [1926, 2] found it still active, provided both enzyme and substrate had been buffered to a suitable pH about 6 before dialysis against distilled water. He suggests that the inactivity in the older experiments was due to the adherence of traces of mineral acid to the soluble starch in spite of dialysis. Moreover, he used a reduction-method, while the older authors generally measured the time needed for the disappearance of the iodine reaction. On the other hand, Sherman, Caldwell, and Adams [1928, 1] found purified pancreatic amylase, which otherwise behaves like ptyalin, to be quite inactive in phosphate solutions between pH 5·7 and 7·7. The general results obtained by Myrbäck are shown in Fig. 30. Michaelis and Pechstein showed that the cation of the salt added was unimportant, Na and K being indistinguishable, Cä and Mg perhaps slightly more favourable than either. Myrbäck found that at a given pH the same activity was shown in the practical absence of salts, or in presence of phosphate, acetate, sulphate, fluoride, or perchlorate (curve I), the other ions investigated having the effects shown. The differences were not due to alterations in the enzyme-substrate affinity.

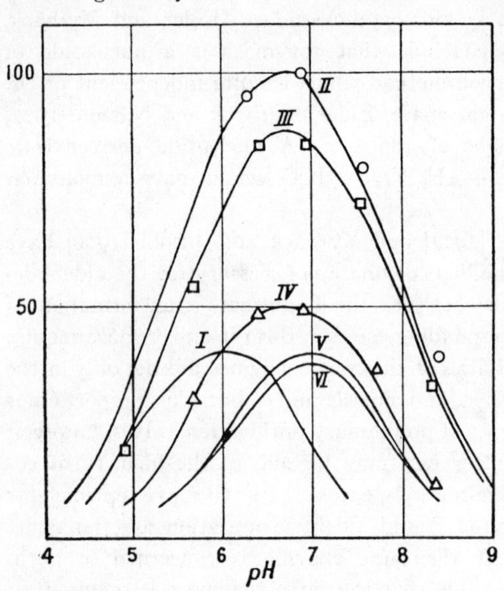

FIG. 30.—Relative velocities of soluble starch hydrolysis by salivary amylase: in presence of phosphate (? + traces of chloride) I, NaCl II, NaBr III, KI IV, $NaNO_3$ V, $KClO_3$ VI, as functions of pH.
[Myrbäck, 1926, 2.]

When NaCl is added to enzyme in 0·1 M phosphate buffer at pH 6·9 the activity is about doubled in ·0005 M NaCl, and increased 3·5 times in ·01 M, after which it remains stationary, falling off again slightly in 0·1 M solution. The values found agree quantitatively with the view that Cl′ forms a compound with salivary amylase whose dissociation constant is 9×10^{-4}, those of the NO_3' and ClO_3' compounds being about five times greater. In mixtures of chloride and nitrate the velocities are intermediate between those in these salts + buffer alone. It is clear that in this case, as with that of lipases, a given substance can only be regarded as an activator under particular circumstances. Thus while chlorides always activate, nitrates activate a phosphate solution at pH > 6·5, but inactivate it in more acid solution, and inactivate amylase + chloride at all pH's. In presence of acetic acid, nitrites slowly destroy the enzyme. In presence of chloride or nitrate, but not fluoride, the rate of destruction is increased about tenfold, which supports the theory of compound formation.

Sherman, Caldwell, and Adams [1928], who find that pancreatic amylase is quite inactive in phosphate and sulphate solutions, obtain the results summarized in Table XVIII.

TABLE XVIII.

Anion.	Cl′.	Br′.	NO_3'.	ClO_3'.	SCN′.	F′.
Maximum velocity	100	77	41	29	29	24
Optimum pH	7·15	7·1	7·1	7·1	6·75	6·75
Optimum concentration of activating ion (M)	·02	·03	·10	·10	·15	·20

The pH-activity curves of pancreatic amylase are exactly the same as those of salivary amylase in presence of phosphate, chloride, nitrate, and chlorate, whence Myrbäck [1926, 2] concludes on their identity, putting down their different stabilities to the presence of accompanying substances. Liveramylase is activated by NaCl [Eadie, 1927]. On the other hand, malt and *Aspergillus* amylases show no special behaviour with salts, although the latter agrees with pancreatic amylase in forming β-maltose from polysaccharides. While many of the older findings as to activation by organic substances were due to changes of pH, there can be little doubt that some behave as genuine activators. Thus Rockwood [1924] found that ·01 M aspartic acid, one of a number of amino compounds, including glycine, aspartic, and hippuric acids, which accelerate ptyalin action, produced a 50 per cent. increase in its activity during an hour's fermentation, and also protected it from decay. But the decay only occurred at the rate of 0·3 per cent. per hour, hence unless it was accelerated several hundred times during digestion (which is unlikely, as the substrate generally protects an enzyme), the effect was not due in the main to protection of the enzyme.

The Complement of the Amylases.

It is well known that when any amylase acts on starch a 100 per cent. yield of maltose is not obtained, but in general about 75 per cent is produced, the remainder of the starch being converted into a dextrin which is only hydrolysed with extreme slowness, if at all. This dextrin appears to be identical with trihexosan [Pringsheim and Beiser, 1924]. In general it arises entirely from the amylopectin of the starch, being also produced from glycogen, but Sjöberg [1924] has obtained it from amylose with a special enzyme preparation. When, however, malt, pancreatic, or salivary amylase is added to suitable protein degradation products, themselves free from amylase or maltase, the digestion of the residual dextrin becomes possible. Complement can be prepared by allowing yeast to autolyse, or by the peptic digestion of egg or blood albumin, myosin, or myogen, but not gelatin, casein, edestin or fibrin-albumose. It is thermostable. On the other hand, proteins and tryptic digests are inactive, and trypsin destroys the complement [Pringsheim and Otto, 1926; Pringsheim and Winter, 1926]. Yeast complement merely catalyses the hydrolysis of residual dextrin, while peptic digests also accelerate the initial stages of starch hydrolysis. Complement-containing solutions combine slowly with maltose, causing a loss of reducing power, while complement-free solutions do not. It is, therefore, probable that dextrins and other substances of low molecular weight are rendered accessible to amylase by combination with protein degradation products. The different end-points found with different preparations in starch hydrolysis probably depend on the amount of complement in the amylase. Where this is small, it all combines with maltose before it has time to act on the dextrin. The same hypothesis explains two anomalous phenomena, the dependence of the end-point on the amount of amylase, and the lack of proportionality between quantity of amylase and reaction velocity.

Enterokinase.

Pancreatic juice when carefully collected without contact with the gut, and suitably prepared pancreatic extracts, have no action on ordinary proteins, though Waldschmidt-Leitz and his colleagues have shown that they will act on certain protein breakdown products and synthetic peptides owing to the presence of a peptidase separable from trypsin (see Chap. VI.), and it has long been known that they act very slowly on casein and fibrin. When an extract of intestinal mucosa is added, they acquire the property of rapidly digesting proteins [Chepovalnikov, 1899]. The process of activation proceeds at first slowly, and then with continually increasing velocity until a maximum proteolytic activity is reached which depends on the amount of pancreatic juice but not on that of enterokinase, as the active substance in intestinal and other tissue extracts is called [Mellanby and Woolley, 1913; Vernon, 1914]. The time taken for activation is, on the other hand, about inversely proportional to the amount of enterokinase added. The first part of the activation has a Q_{10} of about 2, the second, which proceeds a thousand times more rapidly, is almost unaffected by temperature. The increasing velocity of the activation suggested that it was autocatalytic, a fact further proved by Delezenne and Ledebt's [1922] demonstration that at 0° C. pancreatic juice, which only undergoes spontaneous activation with extreme slowness, can be activated by enterokinase. A small fraction of the juice thus activated can then be used to activate more pancreatic juice, and so on indefinitely in series. Vernon further showed the very significant fact that while the amount of trypsin produced by intestinal extract was independent of the amount of enterokinase, nevertheless, when relatively small amounts of active pancreatic extract were added to inactive extract the amounts of trypsin produced were proportional to the amount of active extract. For this and other reasons he believed that activated trypsin solutions contained a special enzyme, deuterase, different from enterokinase. Seth [1924], on the other hand, believed that the activation was due to trypsin itself digesting a protein moiety of trypsinogen. He showed that the optimum pH for activation lies between 6 and 7.

The majority of workers held, as the result of such facts as the above, that enterokinase was an enzyme which produces trypsin from its precursor, trypsinogen, the process being carried to completion by the liberation during the reaction of more kinase, trypsin, or a third body. A few, however, supposed that trypsin was a mere addition product. This latter view is decisively supported by the work of Waldschmidt-Leitz [1924, 1925.] He obtained an exceedingly active preparation of kinase by adsorption and studied its action, not on inactive juice, but on dried preparations of the gland which always possessed some tryptic activity already. He then found that the extra trypsin developed was proportional to the amount of kinase added up to a certain quantity, when a maximum was reached. Activation was complete in 30 minutes at 30° C. Thus 62 mg. of gland developed the following tryptic activities with different amounts of kinase :—

C.c. kinase solution .	·00	·02	·04	·06	·08	·12
Units trypsin . .	·06	·39	·83	1·18	1·31	1·30

In this case ·08 c.c. of enterokinase were needed for complete activation. On doubling and quadrupling the amount of gland, ·16 and ·32 c.c. were needed for complete activation. It follows that trypsinogen and enterokinase unite (or at least co-operate) in definite proportions. The union appears to be fairly stable, since there is no increase in activity when more trypsinogen is added to trypsinogen partly activated by kinase. This is in sharp contrast to the behaviour of cozymase and apozymase. If these latter unite, their compound must be very dissociable. (In Waldschmidt-Leitz' terminology the relatively inactive enzyme is called trypsin, the activated substance trypsin-kinase. I have preferred to keep to the older terminology, especially since trypsinogen has been shown to be quite inactive.) Trypsinogen and kinase can be separated to some extent by adsorption (see Chap. IX.). Bechhold and Keiner [1927] were able to separate them almost completely by ultra-filtration through a suitable collodion membrane. The ultra-filtrate possesses no tryptic activity ; but digests peptone, and hence contains peptidase. The remainder possesses a small residual tryptic activity. On mixing almost the entire original activity is regained. It is concluded that all the kinase and a little of the trypsinogen were retained by the membrane while most of the trypsinogen penetrated it. In neutral solutions kinase is retained by membranes just permeable to hæmoglobin and albumin, in alkaline solutions it permeates more easily. Trypsinogen on the other hand passes through membranes so easily that it may be a crystalloid. Glycerol and ethylene glycol (but not the simple alcohols or mannitol) in concentrations too small for antitryptic action, hinder the action of enterokinase so that more is required for complete activation. So do proteins, for activation almost ceases on the addition of substrates of trypsin. Other inhibitory substances of unknown nature are formed during the autolysis of pancreas.

Grassmann, Dyckerhoff, and Schoenebeck [1930] showed that cystine in relatively large amounts may replace kinase as activator of trypsinogen. On the other hand, cysteine and pyrophosphate inhibit tryptic activity when small amounts of kinase are present. But the inhibition is largely prevented by excess of kinase. Hence the inhibitors must compete with kinase for trypsinogen, the trypsinogen-inhibitor compound being relatively inactive.

A glycerol extract of pancreas, which gradually undergoes activation when left standing, can be separated by adsorption on alumina at pH 4·7. The adsorbate includes not only erepsin, but a substance which gradually develops into kinase, while the residue includes the trypsinogen [Waldschmidt-Leitz and Harteneck, 1925]. There is, therefore, a prokinase in the pancreas, and it is reasonable to suppose that the conversion of this substance into enterokinase is the limiting factor during the first stage of activation under the conditions studied by such workers as Mellanby and Woolley, the combination of the kinase with trypsinogen accounting for the second stage (Vernon's deuterase action). Waldschmidt-Leitz and Harteneck suggest that prokinase is activated by erepsin. It seems possible, however, that traces of trypsin may be responsible. It is known that the activation of pancreatic juice after the addition of small amounts of kinase is an autocatalytic process, and since enterokinase has been shown not to be a catalyst it is simplest to suppose (with Seth and others) that the catalyst concerned is trypsin, but that it acts, not on trypsinogen, but on prokinase. So that two successive reactions are involved, namely,

Prokinase → kinase (catalysed by trypsin)

kinase + trypsinogen → trypsin (uncatalysed).

It can be shown that when small quantities of kinase are added to a trypsinogen + prokinase solution three stages should occur, a short period of induction during which the added kinase is combining with trypsinogen, a period which may be of considerable length during which the amounts of kinase and trypsin increase exponentially, and a final period during which the velocity does not increase, owing to the exhaustion of either trypsinogen or prokinase. This hypothesis appears to fit the facts so far observed, but it is clear that until the activation of pancreatic juice by purified kinase and trypsin has been quantitatively investigated, the matter will not be settled.

The Activation of Plant Proteases.

Papain (i.e. the proteolytic enzyme of *Carica papaya* juice) attacks proteins near their isoelectric point, particularly denatured proteins such as gelatin. In presence of HCN the activity as measured by amino N production is increased two to three times, and the specificity widened, as native proteins, peptones, and tripeptides, such as leucyl-glycyl-leucine, though not dipeptides, are now attacked [Willstätter and Grassmann, 1924]. The pH optimum is unaffected. It is not clear whether the activity on proteins is really increased, or whether the apparent increase is merely due to a completer hydrolysis of the split products. When HCN is added to papain the process of activation is not instantaneous. The maximum activity is reached after a little more than an hour at 40° C., after which it falls off by about 5 per cent., perhaps because a certain mixture of papain and papain-HCN is more effective than either alone, the free enzyme being more efficient as a catalyst of certain reactions. The activation is fully reversible by a vacuum, and can then be repeated. Papain is also activated by H_2S, but several nitriles, amygdalin, and various other organic substances have no effect. HCN is effective in concentrations of 10^{-5} M and over. It has been suggested that the activation is due to the combination of the cyanide or sulphide with a heavy metal which inactivates the papain. Both HCN and H_2S will, for example, activate saccharase which has been poisoned by Ag salts (see Chap. VIII.). As against this view no case is known except possibly in the case of bacterial dehydrogenases (see Chap. VI.) where a single enzyme can be poisoned so as to abolish its activity for certain substrates only. And all attempts to fractionate papain have failed. Willstätter and Grassmann found that activation was quantitatively the same whether crude material or an enzyme purified by adsorption or alcohol precipitation was used. The chemical process underlying activation is therefore quite uncertain. Pineapple protease behaves similarly to papain with HCN and H_2S, while Cucurbita protease is inhibited [Willstätter, Grassmann, and Ambros, 1926]. Yeast protease behaves like papain, the polypeptidase being inhibited [Grassmann and Dyckerhoff, 1928, 3]. Cysteine, but not cystine, has the same effects as HCN [Grassmann, Dyckerhoff, and Schoenebeck, 1930]. When crude solutions of these proteases are allowed to stand, or in certain fresh preparations, they behave as if activated by HCN. The activator "phytokinase," is unstable at pH 5 [Grassmann and Dyckerhoff, 1928].[1]

The "Activator" of Myozymase.

Extracts of frog's muscle produce lactic acid rapidly from starch, glycogen and trihexosan; slowly from the sugars. Thus Meyerhof [1927] found that galactose and saccharose were unaffected, glucose

[1] Waldschmidt-Leitz, Purr, and Balls [1930] have shown that the activator of animal kathepsin is glutathione.

yielded lactic acid at 20 per cent. to 40 per cent. of the rate of starch, while fructose, mannose, maltose, and amylobiose were fermented at intermediate rates. Some extracts of rabbit's muscle at first fermented glucose as rapidly as starch, but the capacity for doing so fell off on standing more rapidly than that for fermenting starch. When, however, a substance obtained by the alcoholic precipitation of autolysed yeast is added, the velocity of lactic acid production is greatly increased, and in certain preparations it may be at first greater than that from starch or glycogen. However, whereas in the case of polysaccharides only a small proportion of the glucose is esterified with phosphoric acid, in hexose fermentation with the activator half of it passes into fructose diphosphate, as in alcoholic fermentation. As soon as the phosphate is used up the reaction therefore slows down, being limited by the velocity of phosphatase action. The activator, unlike insulin, cozymase, and amylase complement, is destroyed in 5 minutes at 50° C., and is only stable at ordinary temperatures in neutral solution. It seems probable, therefore, that the activator is an enzyme, and a member of the zymase complex. Against this view is the fact that it has no effect on sugar except in presence of the muscle enzyme.

Respiratory Coenzymes.

When chopped muscle or acetone yeast is washed, it loses its capacity for reducing methylene blue, and the washings do not possess this faculty. On the other hand, the two together do so. Battelli and Stern [1911] ascribed this to a coenzyme, pnein, in the washings. Meyerhof [1918, 1] observed similar phenomena in killed yeast. Part of this action is due to the presence of hydrogen donators such as succinic acid in the washings, part to glutathione. Glutathione, besides being autoxidizable, catalyses the oxidation of unsaturated fatty acids and proteins [Hopkins, 1928]. These are, however, not enzymatic processes, and glutathione does not restore the "coenzymatic" power to a tissue extract which has lost it by standing aerobically with chopped muscle, nor has it been shown to act as a coenzyme in any system so far described. Holden [1924] regards the "atmungskörper" (respiratory substance) as consisting entirely of oxidizable substances, since it disappears on standing in presence of air with chopped muscle, though not without it, but he did not discover whether it could be destroyed anaerobically. Meyerhof [1918, 2], however, finds that the respiratory substance, or a large fraction of it, agrees with cozymase in its dialysability, moderate thermostability, precipitability by alcohol and lead acetate, and destruction by alkalies. These are not properties of hydrogen donators known to be present in the cell. Other similarities are noted on page 137, but the question is not yet settled.

Urease.

Onodera [1915] found that on dialysis urease lost its activity, which is not restored, except to a very slight extent, on adding the dialysate to the contents of the parchment paper. On the other hand, a very small amount of fresh urease restores nearly full activity to the inactivated enzyme. In a similar way the activity of urease heated to 80° C. for an hour is restored by a small amount of fresh

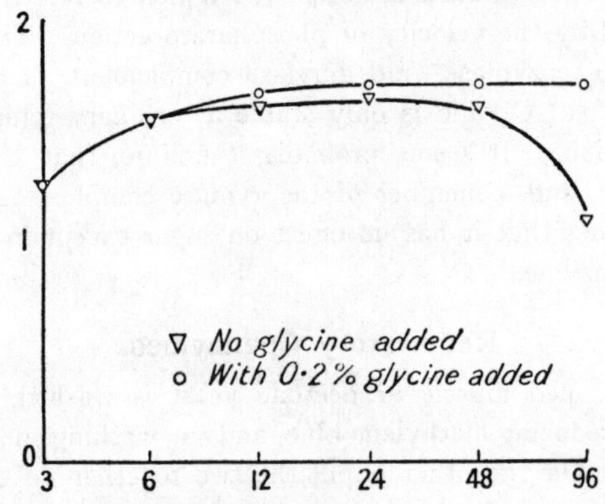

FIG. 31.—Effect of added glycine on urease hydrolysis of urea at varying concentrations. Abscissa, urea concentration, milligrams in 5 c.c. (logarithmic scale). Ordinates, velocities of urea hydrolysis. [After Kato, 1923, 1.]

enzyme. Onodera regarded the effect of dialysis and heating as the loss of a coenzyme. As the latter could not be found in the dialysate it may equally well be supposed that the effect of dialysis and heating was to destroy a colloidal substance. Lövgren [1927] confirmed his results, using buffers. Kato [1923] found that these effects were functions of the urea concentration, and that the stable component of urease could be replaced by glycine. The effect of either of these substances is to increase the action of urease in strong urea solutions (Fig. 31). He found, however, that the effect was quantitatively very different on different preparations, and it is not necessarily those samples which fall off most in strong urea solutions which undergo the greatest activation in moderate concentrations. In a sufficiently strong solution it is clear from Fig. 32 that the amount of heat-stable substance determines the rate of action, since the rate is just doubled by the addition of an equal amount of boiled enzyme. The heat-stable substance is very much more active than glycine. The effect of adding more glycine gradually falls off as the amount increases. and ultimately the amount of enzyme becomes the limiting factor. Rockwood and Husa [1923] tested the action of forty-eight compounds in ·001 M concentration on the hydrolysis by urease of urea in 0·1 M solutions. The effects were, therefore, relatively small. All α-amino-acids, including glucosaminic acid, had a positive effect, ranging from + 9 per cent. (valine) to + 62 per cent. (histidine). β- and γ-amino-acids and amino-benzoic acids had no effect, nor did a number of other substances, except where they inhibited. Witte's peptone,

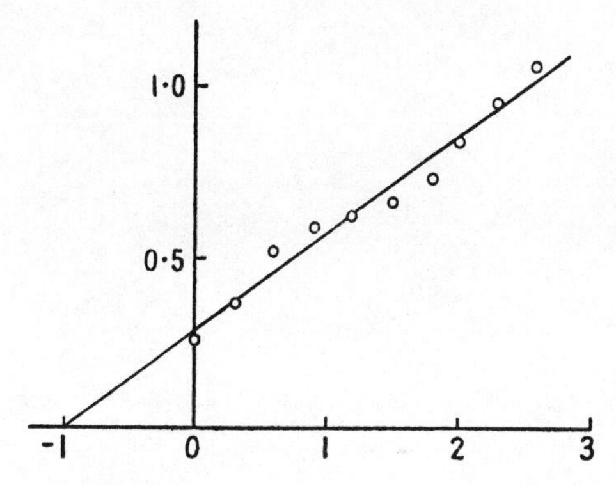

Fig. 32.—Effect of addition of boiled urease on the rate of hydrolysis of 3 per cent. urea by urease. Abscissa, c.c. of boiled urease solution added to 1 c.c. unboiled solution. Ordinate, rate of hydrolysis. The prolongation to the left shows that the heat-stable factor determined the rate of hydrolysis by the unboiled enzyme. [After Kato, 1923, 2.]

however, was active. The only exceptions were substituted amino-acids, namely, hippuric acid and histidine methyl ester, which were as active as their parent compounds. Changes of pH were excluded. Rockwood [1924] showed that these compounds protected urease from slow inactivation on standing in solution, but also had a specific effect. This also follows from many of Kato's protocols, since he found that during the first two hours the velocity of urea hydrolysis, either with or without glycine, was absolutely constant.

Kato found that, when the hydrolysis was carried out at 15° C., but not at higher temperatures, the action was more accelerated if the glycine and urea were left overnight in the ice chest than if they were mixed at once, and concluded that they formed a compound.

I have, however, been unable to obtain any evidence of its existence by cryoscopy.

Taubmann [1925] showed that, in experiments lasting for thirty-six hours or over, but not in short experiments, gum-arabic or starch had a favourable influence on urease action. The time factor suggests that the effect is protective. Sumner and Hand [1928] found this effect enormously enhanced with highly purified urease and attributes it to protective action against traces of lead. It is fairly clear that in the case of amino-acids protective and coenzyme action are both present.

CHAPTER VIII.

THE POISONING OF ENZYMES.

APART from physical factors and changes of pH, a very large number of chemical changes may depress the action of an enzyme. The change may, of course, be reversible or irreversible, but in many cases it is not known which of these alternatives is true. It is not, therefore, possible as yet to divide the actions of poisons according to their reversibility, which is theoretically desirable.

The following classification of enzyme inhibitors apart from substances related to the substrate, which are discussed in Chapter III., and adsorbents, discussed in Chapter IX., may be made provisionally:

1. Substances combining with the enzyme as an acid, e.g. heavy metals.

2. Substances combining with the enzyme as a base, e.g. acid protein precipitants.

3. Oxidizing agents.

4. Amine reagents, e.g. HCHO, HNO_2.

5. Aldehyde reagents.

6. HCN, H_2S, $H_4P_2O_7$, NO.

7. Surface-active substances, e.g. substituted urethanes, and ions influencing colloidal behaviour.

8. Other crystalloidal substances whose mode of action is obscure.

9. Colloidal and semicolloidal inhibitors, including immune bodies.

Heavy Metals.

Enzymes are usually inactivated by salts of the heavy metals, such as silver and mercury, in very small concentrations. The action of $\dot{Ag}$ ions on purified yeast saccharase has been very fully investigated by Myrbäck [1926]. Fig. 33 shows the rate of hydrolysis of cane sugar at various pH's in concentrations of $AgNO_3$ ranging from 0 to 10^{-4} M. Acetate buffers were used, as silver phosphate is very insoluble. The inactivation is, except in very impure preparations, unaffected by the concentration of cane sugar, and fully reversible by H_2S [Euler and Svanberg, 1920]. The results can be explained on the hypothesis that the silver ions combine with the anions of saccharase, but not with the neutral molecules or cations. More precisely the silver combines with that particular acidic group of the saccharase whose dissociation determines its efficiency as a catalyst. On Bjerrum's theory of ampholytes this group, with pK = 6·5 might be an amino group. The affinity constant of this union must be unaffected by the union of the enzyme with sugar. We have then the two equilibria

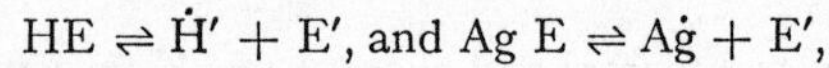

$$\mathrm{HE} \rightleftharpoons \dot{\mathrm{H}}' + \mathrm{E}', \text{ and Ag E} \rightleftharpoons \dot{\mathrm{Ag}} + \mathrm{E}',$$

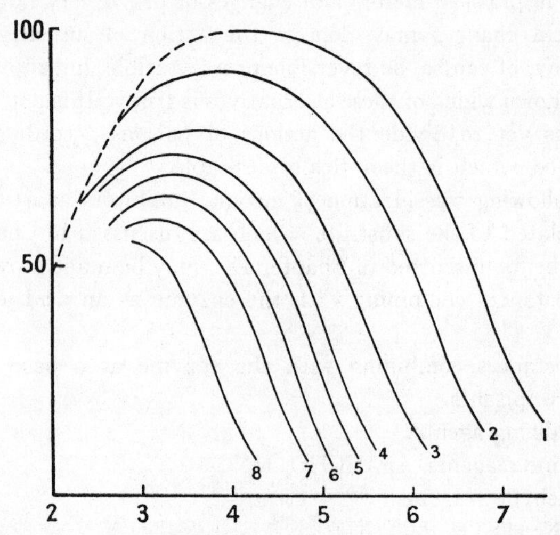

FIG. 33.—Poisoning of yeast saccharase by different $AgNO_3$ concentrations as a function of pH. Abscissa, pH. Ordinates, initial velocities of hydrolysis of 10 per cent. sucrose. $AgNO_3$ concentrations :—
1. 0; 2. 5×10^{-7}M; 3. 10^{-6}M; 4. 2×10^{-6}M; 5. 4×10^{-6}M; 6. 10^{-5}M; 7. 2×10^{-5}M; 8. 10^{-4}M. [After Myrbäck, 1926, 1.]

where only HE molecules are active. Hence

$$[\dot{\mathrm{H}}]\,[\mathrm{E}'] = \mathrm{K}_a[\mathrm{HE}],\ [\dot{\mathrm{Ag}}]\,[\mathrm{E}'] = \mathrm{K}_{\mathrm{Ag}}\,[\mathrm{AgE}],$$

where K_a is the acid dissociation constant of the enzyme on the Michaelis theory, K_{Ag} the dissociation constant of its silver compound. If e be the total molecular concentration of the enzyme, x the activity in comparison with that at optimum pH in the absence of silver, and y the total molecular concentration of silver,

$$\therefore \left(1 + \frac{\mathrm{K}_a}{c\mathrm{H}}\right) ex^2 + \left(y + \mathrm{K}_{\mathrm{Ag}} - e + \mathrm{K}_{\mathrm{Ag}}\frac{c\mathrm{H}}{\mathrm{K}_a}\right) x - \mathrm{K}_{\mathrm{Ag}}\frac{c\mathrm{H}}{\mathrm{K}_a} = 0.$$

When y is large compared with e or K_{Ag}, $x = \dfrac{\mathrm{K}_{\mathrm{Ag}} \cdot c\mathrm{H}}{y \cdot \mathrm{K}_a}$ whence K_{Ag}

can be calculated, and is found to be $4 \cdot 10^{-8}$, as compared with $K_a = 3 \cdot 10^{-7}$. In the above equation no allowance is made for the basic properties of saccharase, hence it only holds for the alkaline side of the isoelectric point. The most important feature of the above equation is that it enables us to calculate e, the molecular concentration of enzyme, which in three particular solutions was of the order of 10^{-6}, giving a combining weight for the enzyme of about 5000. As the enzyme was certainly not pure, this value is only an indication. But it is a striking fact that the amounts of silver needed to reduce the enzymic activity to half were almost unaffected by twelve-fold purification of the enzyme. Hence any other silver-combining bodies remain with the enzyme during purification. Gold and mercury are about as poisonous as silver; thorium, copper, lead, zinc, and cadmium about $\frac{1}{10000}$ as poisonous; while Al, Cr, Mn, Ni, and Co and Be are still less toxic. We have clearly to deal with specific chemical action and not with colloidal behaviour. The effect of Hg is of a different nature to that of Ag, as will be seen later. The silver saccharase compound is soluble, but more easily adsorbed by alumina than the free enzyme.

The above inactivation is very rapid. Impure saccharase preparations whose activity has been reduced to about 50 per cent. by silver salts recover 80 per cent. to 90 per cent of their original activity in the course of about six hours, but do not recover completely. This is undoubtedly due to a transference of some of the silver to other combinations, since it does not occur with purified enzyme [Euler and Myrbäck, 1923]. Olsson [1925] found that metal-poisoned amylases were more readily inactivated by heat than the free enzyme.

Other enzymes behave in the same way. It is an important fact that whereas Michaelis and Pechstein [1914] found that at its optimum pH of 6·0 salivary amylase in acetate buffer migrates to the anode, yet Myrbäck's [1926, 2] work shows that it behaves both to anions and cations as if its isoelectric point coincided with its optimum pH whether in presence or absence of NaCl. At any rate the acid and basic groups whose combination affects the activity of the enzyme behave as if this were so. The following enzymes are poisoned by one or more heavy metals:—

Pancreatic, liver and *Ricinus* lipases, yeast invertase, salivary, pancreatic and malt amylases, urease, pepsin, trypsin, and yeast protease.

But the relative effects vary greatly.

Kastle [1906] found that $HgCl_2$ and $CuSO_4$ were much more poisonous to liver lipase than $AgNO_3$, which is barely poisonous, whereas malt amylase is inactivated by saturated AgCl solution [Olsson, 1921]! The surface catalysts of oxidation and reduction in *Bacillus coli* can be poisoned with copper salts, and the action reversed by H_2S. Reversibility is more complete for some substrate activations than for others. Copper does not protect the oxidases from toluene, aniline, phenol, or methylene blue poisoning [Quastel and Wooldridge, 1927, 1, 2].

Acid Protein Precipitants.

Yeast saccharase is inactivated by dilute picric, phosphotungstic and silicotungstic acids, the inactivation being almost confined to the acid side of the optimum, and inappreciable in solutions more alkaline than pH 5. But, unlike the effect of silver, this inactivation is much less marked in strong than in weak sucrose solutions, i.e. it is competitive [Myrbäck, 1926, 1]. If we accept the argument that the acids combine with a basic, presumably amino group of the enzyme (and it is difficult to see what else phosphotungstic acid could do), it follows that cane sugar combines with the same group, or at least with a group so near to it that steric hindrance renders simultaneous union of the enzyme with acid and sugar impossible.

Rona and Bach [1921] found that the inhibition of saccharase by nitrophenols is a relatively slow and irreversible process. Picric, phosphotungstic, and tannic acids behave in the same way with regard to salivary, pancreatic, and malt amylases, except that in the case of tannic acid the inactivation is sometimes progressive. Nothing is known in most cases as to the influence of substrate concentration.

Oxidation.

All enzymes so far investigated are destroyed by relatively mild oxidizing agents. Chlorine [Myrbäck, 1926, 1] and bromine [Euler and Josephson, 1923, 2] inactivate yeast saccharase roughly in proportion to the amount added, except that a little may be taken up by reducing substances before any effect is produced. Chlorine inactivation is complete within thirty seconds. Iodine reduces the activity at once to 60 per cent. of the original. Subsequently it declines gradually to zero, the rate being slowest at a pH near the optimum. Myrbäck believes that an iodo-saccharase is formed, and in presence of excess of iodine slowly destroyed. Iodo-saccharase has the same pH-activity curve and Michaelis constant as the untreated enzyme, in contra-distinction to enzyme treated with heavy metals or acid protein precipitants. Euler and Josephson found that 1 mol. Br_2 inactivated 4000 to 5000 gms. of their purified saccharase, 1 mol. I_2 halved the activity of 20,000 gms. Saccharase is not protected against iodine by cane sugar. Thiosulphate does not reactivate the completely inactive enzyme, but it wholly inactivates iodo-saccharase, while having no effect on the normal enzyme. Saccharase is slowly inactivated by OsO_4. A number of other enzymes are inactivated by halogens.

Hydrogen peroxide in moderate concentrations inactivates a number of enzymes, including urease, fig protease, certain rennins, peroxidase (see Chap. III.), xanthine oxidase. But small quantities activate xanthine oxidase (Chap. IV.). Halogens inactivate amylases (there are differences in their susceptibility to I_2) certain rennins, catalase, and the surface oxidase action of "resting" *Bacillus coli.* In this last case there is a decided difference in the order in which different catalysts were inhibited by H_2O_2, $KMnO_4$, KNO_2 and the halogens. Partial reversal of certain inhibitions by reducing agents was observed in the case of $KMnO_4$ and KNO_2 [Quastel and Wooldridge, 1927, 1, 2].

McCance [1925, 1] found that an enzyme producing urea from an unknown soluble precursor (not arginine) in tissue autolysis was inactivated reversibly by O_2, which has no effect either on the production of soluble amino nitrogen or of NH_3. The inactivation is not complete.

Amine Reagents.

Among the enzyme poisons are included aldehydes, HNO_2, and $\ddot{Hg}$, which react with amino groups. Yeast saccharase is slowly inactivated by nitrous acid at pH less than 5, and the rate is proportional to the proportion of nitrite present as HNO_2. If the substrate combines with the amino group, it should protect it from diazotization. Myrbäck [1926, 1] found that cane sugar, glucose, and fructose, do so protect it, while lactose and maltose, which do not appear to unite with the enzyme, do not protect it.

Mercuric chloride inactivates yeast saccharase, predominantly in alkaline solutions. The inactivation is fully reversible by H_2S and NaCN. Nevertheless, the pH-activity curves in presence of $HgCl_2$ are very different from those obtained with silver, and cannot be represented by the same equation. For a given degree of inactivation on the alkaline side of the optimum there is relatively far more inactivation on the acid side. Moreover, the enzyme is protected by strong cane-sugar solutions. Finally mercury protects the enzyme to some extent against inactivation by HNO_2, while silver does not. Myrbäck [1926, 1] therefore believes that the metal forms a complex with the active amino group. The affinity is very great, half inactivation at pH more alkaline than 5 occurring in $5 \cdot 10^{-7}$ M $HgCl_2$. Mercurous and mercuric salts are equally toxic, which further supports the view that the reaction is not ionic. On the other hand, salivary amylase appears to behave in the same way with Hg salts as with those of other metals [Myrbäck, 1926, 2].

Aldehydes poison a number of enzymes. In the case of yeast saccharase [Myrbäck, 1926, 1] the enzyme is gradually inactivated by formaldehyde, inactivation being slower at the optimum pH than at higher or lower acidities. There is no definite evidence of combination with an amino group. As mentioned in Chapter II., formaldehyde poisons trypsin, but not pepsin. This may be due to the fact that trypsin, but not pepsin, combines with its substrate through an amino group. It may, however, merely be due to the fact that pepsin, which, when pure, is not an ampholyte, possesses no amino groups. *Ricinus* lipase, maltase, emulsin, and urease are also inactivated by formaldehyde, and the surface oxidase actions of *Bacillus coli* by formaldehyde and acetaldehyde.

Aldehyde Reagents.

Myrbäck [1926, 1] studied the inhibition of yeast saccharase by aniline and a variety of substituted anilines. The inhibition, which is completely reversible, is generally much more marked in alkaline than in acid solutions, just as is the formation of compounds with aldehyde, owing to the fact that in alkaline solutions the bases are largely dissociated. On the other hand, *o*-, *m*-, and *p*-chloraniline, which are very weak bases, have a poisonous effect independent of pH provided the latter exceeds 3, except that inhibition is increased in the neighbourhood of pH 7. The effect of varying the concentration of aniline agrees quantitatively with the view that a compound is formed with the enzyme. The dissociation constants of the compounds so formed with sixteen bases were calculated, and were found to show a general parallelism with their basic dissociation constants, except that the para compounds were relatively much less toxic than the ortho and meta. However, owing to the effect of position on the basic dissociation constant, the order of toxicity was always meta, para, ortho. While the dissociation constants of the enzyme-base compounds vary from 0·3 (*o*-aminobenzoic acid) to ·0001 (*m*-toluidine), the product of the two dissociation constants does not vary more than tenfold within either ortho, meta, or para-compounds. The degree of inhibition by these bases is quite independent of the substrate concentration.

Myrbäck now determined the dissociation constants of a number of formaldehyde-base compounds, and found no parallelism with those of the saccharase-base compounds. However, when the dissociation of base-lactose compounds was determined by allowing the lactose to protect the enzyme from the bases, it was found that, as in the case of the enzyme, the order of affinity is always meta, para, ortho, but the dependence on the strength of the base was less. Myrbäck came to the conclusion that the aldehyde group of saccharase is part of a carbohydrate residue.

Phenyl-hydrazine and amino-guanidine inhibit strongly, especially in alkaline solutions, hydroxylamine slightly, and semicarbazide not at all. Cyanides have a very small effect, and sulphites none. All these facts suggest that if an aldehyde group is present, it is part of an aldose group.

Saccharase is also reversibly inactivated by a number of alkaloids. [Rona and Bloch [1921], Rona, v. Eweyk, and Tennenbaum [1924].] Here also the effect was greater in neutral than in acid solutions, and the molar concentrations producing half inhibition were of the same order of magnitude as in the case of the more basic substituted anilines. But in many cases it is clear that no compounds of the Schiff's base type could be formed. Thus methylatropine bromide, in which the only N atom is pentavalent, produced half inhibition in ·0001 M solution. It is clear, therefore, that organic bases may act in a manner different to that postulated by Myrbäck, and some doubt is thrown on his conclusions. Moreover, the effect of varying the alkaloid concentration shows that the enzyme-alkaloid union is not a simple combination, but an adsorption. Purification of the enzyme had no effect on the phenomena.

Phenylhydrazine, hydroxylamine, semicarbazide, sodium sulphite and sodium cyanide all poison malt amylase [Olsson, 1925], which points to the presence of an aldehyde group, and catalase and peroxidase are poisoned by hydrazine and hydroxylamine [Loew, 1901], [Wieland and Sutter, 1928]. On the other hand, Rona [1920] in unbuffered or incompletely buffered solutions found no substantial effects with the following: —

Yeast saccharase + phenylhydrazine.

Maltase + $Na_2 SO_3$.

Emulsin + Na_2SO_3 or KCN.

Malt amylase + Na_2SO_3.

Pepsin + Na_2SO_3, NH_2OH, HCl, or benzene-sulph-hydroxamic acid.

Trypsin + Na_2SO_3, KCN, or phenylhydrazine.

The absence of effect of phenylhydrazine on saccharase suggests that the work should be repeated.

Aniline, phenylhydrazine, semicarbazide, and amino-guanidine inactivate the surface oxidase activity of *Bacillus coli* [Quastel and Wooldridge, 1927, 2], but rather large concentrations are required.

HCN, H_2S, $H_4P_2O_7$, and NO.

Cyanides do not in general poison the hydrolytic enzymes, except in concentrations of the order of 1 per cent. However, erepsin [Euler and Josephson, 1926] and the yeast peptidases [Grassmann and Dyckerhoff, 1928, 1] are inhibited by smaller concentrations of HCN and H_2S. The yeast peptidases are also inhibited by cysteine, but not cystine [Grassmann, Dyckerhoff, and Schoenebeck, 1930]. On the other hand, even in very low concentrations HCN poisons zymase, plant peroxidase, phenolases such as laccase, tyrosinase and catalase, besides succinoxidase and surfaces which catalyse oxidation. However, Fleisch [1924] and Szent-Györgi [1924] found that while the reduction of O_2 by succinoxidase + succinic acid is inhibited by cyanides, that of methylene blue is not. Xanthine oxidase, on the other hand, is not inhibited by cyanides either in its reduction of methylene blue or of O_2 [Dixon and Thurlow, 1925]. Lactic dehydrogenase, which reduces methylene blue, but not O_2, is also immune [Stephenson, 1928]. Most other oxidases so far investigated are inhibited by HCN; on the other hand, this substance does not completely inhibit oxidation by most intact cells [Dixon and Elliott, 1929], and in the case of chlorella [Warburg, 1919] even accelerates it. The poisoning is usually reversible, though in the case of certain surface oxidations it is irreversible [Quastel and Wooldridge, 1927]. In the case of the whole yeast cell the poisoning is independent of the O_2 concentration [Warburg, 1927], and unpublished work of the author shows that this is approximately true for moths. Warburg [1928] ascribes all cyanide poisoning to the formation of compounds with heavy metals, and this view is supported by the behaviour of non-enzymatic catalysts of oxidation, such as cysteine and impure reduced glutathione. These are readily oxidized, either by O_2 or methylene blue. When cyanides are added, both these processes are very greatly slowed down. When the solution is carefully freed from iron, the oxidation is extremely slow, and the effect of cyanide is greatly reduced [Harrison, 1927; Toda, 1926]. On the other hand, Warburg [1928] has held that the function of the heavy metals is to activate O_2, and regards cyanide poisoning as an interference with this activation. To the writer it would appear that the best evidence that O_2 is being activated by a special catalyst is the poisoning of the system by CO, competitively with O_2. The fact that HCN poisoning is non-competitive with O_2 does not, of course, prove that it is not an inhibition of an oxygen activator; on the other hand, while all systems poisoned by CO are also poisoned by HCN the converse is not true, and some anaerobic oxidation-reductions are poisoned by HCN. If, therefore, in such systems as the indophenol and polyphenol oxidases studied by Keilin [1929] the reducing substrates as well as O_2 are being activated, which is not impossible, no really conclusive evidence remains that the process inhibited by HCN is the activation of O_2. The best evidence for this hypothesis appears to be the poisoning by HCN of the reduction of O_2, but not of methylene blue, by succinic acid and succinoxidase. It is certain, in any case, that O_2 activation is not the only reaction poisoned by HCN, since, e.g. peroxidase and zymase are affected by it.

It has been suggested that, in some cases HCN may poison enzymes by combining with a carbonyl group rather than a heavy metal. However, both in the case of oxidases and peptidases, H_2S is a poison of about equal strength with HCN, which supports the heavy metal hypothesis. $H_4P_2O_7$ has a similarly poisonous effect on cell respiration, though Dixon and Elliott [1929] showed that it does not poison exactly the same set of catalysts as HCN. It also inhibits the yeast peptidases [Grassmann, Dyckerhoff, and Schoenebeck, 1930], and in low kinase concentrations, trypsin (see p. 142). To sum up, HCN and H_2S poison a number of catalyses, probably by heavy metals in organic combination. If this is correct, not only many oxidizing-reducing enzymes, but also some peptidases must contain heavy metals in their molecules.

Warburg [1927] found that NO poisons alcoholic fermentation by yeast, which is not affected by CO, and only mildly by HCN and H_2S (it is obvious that its effect on respiration cannot be tested).

He attributes this poisoning also to combination with a heavy metal compound which is catalysing some stage of the fermentation.

Surface-active Substances.

Various enzymes are reversibly inhibited both by substances such as the substituted ureas and urethanes which possess a narcotic action and are adsorbed by various surfaces from aqueous solution, and by salt mixtures containing ions, e.g. Cä, which are adsorbed by proteins. Crude yeast saccharase shows both these effects [Meyerhof, 1914; Neuschloss, 1920]. But Schürmeyer [1925] found that neither of these effects occurred with saccharase so highly purified as to be practically protein-free. When serum albumin is added, it becomes moderately sensitive to such a narcotic as phenyl-urea, when serum para-globulin is added, much more so. For example, 0·3 per cent. phenyl-urea produced 0·0 per cent. to 0·2 per cent. inhibition with pure enzyme, 12 per cent. and 24 per cent. with crude preparations, 6 per cent. with pure enzyme + ·013 per cent. albumin, and 34 per cent. with pure enzyme + ·014 per cent. globulin. A further tenfold increase in the protein concentration had little effect in increasing the inhibition. The crude preparation shows inhibition in varying concentrations of alcohols and substituted urethanes in agreement with the view that they are adsorbed by it, and the intensity of the effect for a given concentration increases with molecular weight in accordance with Traube's law. Inhibition is slightly greater in weak than in strong cane-sugar solutions, but whereas Meyerhof found that 1·5 M methyl-urethane produced 30 per cent. inhibition in 9 per cent. sucrose, ·75 M produced the same in 0·3 per cent. Hence the inhibition is very nearly non-competitive. The fact that the poisoning is due to adsorption was taken to show that saccharase acts by adsorption of cane sugar, and the fact that it does not occur in pure preparations is equally cogent evidence against this view.

Schürmeyer (who adjusted his pH with HCl and NaOH) found that the chlorides of Na, K, Ca, and Mg had no effect on purified saccharase, or on saccharase + proteins or lecithin, provided the pH was kept on the acid side of the isoelectric points of the added colloids. When albumin or gelatin was added Ca and Mg salts depressed the activity on the alkaline side of the isoelectric points, but there was no definite antagonism between them and Na and K. But in presence of globulin or lecithin the effects of univalent and bivalent cations were wholly or partially neutralized at an appropriate ratio (Fig. 34). Bä and Lä ions also antagonized the univalent ions.

Similar inhibitions have been obtained with the enzymes concerned in alcoholic fermentation and in oxidation. As they are parallel to the effects on the living cell, they no doubt explain these latter. However, until Schürmeyer's work has been repeated with other purified enzymes, it would seem that they do not necessarily throw any light on the nature of enzyme action; although they are, of course, important in any consideration of the physiology of enzymes, and tend to show that they are united with other colloids in the living cell.

Velluz [1927] studied the effects of fatty acids on the activity of

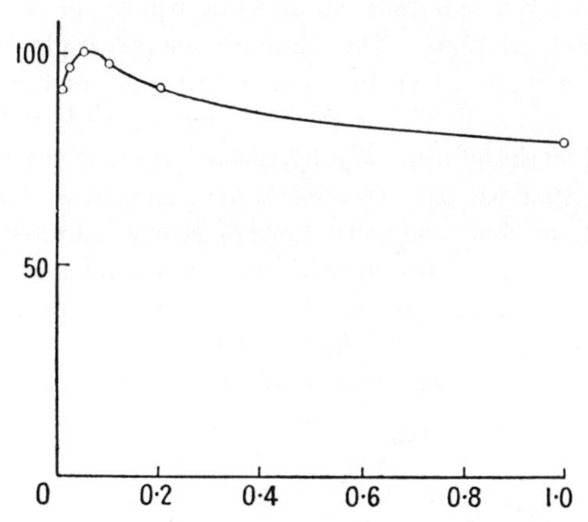

Fig. 34.—Activities of purified saccharase + globulin in presence of varying Na : Ca ratios. Abscissa, ratio of [Na] : [Ca] in $\frac{M}{2}$ salt mixture. Ordinate, rate of sucrose hydrolysis as percentage of constant rate of hydrolysis by protein-free enzyme in the same mixtures. [After Schürmeyer, 1925.]

pepsin and urease. In the former case he found an increase in inhibitory power on ascending the series from caproic ($C_6H_{12}O_2$) to stearic ($C_{18}H_{36}O_2$) acid. Unsaturated acids are more effective than saturated in the case of both pepsin and urease, and in the latter case inhibition appears to be partly competitive with the substrate. Some colloids, such as gum-arabic, counteract the inhibition, probably by adsorbing some of the fatty acid.

Other Crystalloids.

One of the most universal of enzyme poisons is fluoride. The enzymes affected by it include gastric, pancreatic, and *Ricinus* lipases, malt amylase, urease, pepsin, myozymase, and some dehydrogenases. The relative effects, however, differ greatly. Thus Rona and Pavlovič [1923] found that NaF inhibited liver and serum lipase strongly, pancreatic weakly. It seems possible that this action is merely due to its position in the Hofmeister series, which enables it to affect the colloidal properties of the enzymes or enzyme-colloid aggregates. This view is supported by the finding of Meyer [1928] that purified myozymase is less affected by fluoride than the crude enzyme. If the above explanation is correct, fluorides fall in with the substances considered in the last paragraph.

A number of substances of no great chemical activity, such as the alcohols, inhibit in relatively large concentrations. The behaviour of different enzymes varies, some being still active in fairly strong alcohol solutions. The data do not generally permit of a distinction between reversible and irreversible inactivation. The necessity of such a distinction is shown by the work of Hudson and Paine [1910] on the influence of ethyl alcohol on crude yeast saccharase. They found that whereas irreversible activation was most rapid in 50 per cent. alcohol, and very slow in strong solutions, reversible inactivation increased steadily with the amount of alcohol, and this has since been shown (see p. 30) to be due to dilution of H_2O. Willstätter and Racke [1921] found that the rate of inactivation by alcohol and acetone was greatly increased by purification of the enzyme. It is not yet possible to arrive at any general conclusions as to the mode of action of these bodies on enzymes.

A large amount of work has been done, especially by Rona and his colleagues, on the action on enzymes of alkaloids and organic arsenic compounds. On yeast saccharase (see p. 155) they found that the inhibition was proportional to a fractional power of the alkaloid concentration, i.e. assuming that the inhibition was proportional to the amount of alkaloid adsorbed, the adsorption isotherm followed Freundlich's equation over a wide range, though this equation cannot hold in very strong alkaloid solutions. Only the free base is active, so that inhibition increases with pH over a certain range. It is not known whether purified saccharase is affected. Particular attention was paid to lipases from serum, liver, stomach, pancreas, and other organs, and of various mammals [Rona and Bach, 1920; Rona and Reinecke, 1920; Rona and Pavlovič, 1922; Rona and Petow, 1924; Rona and Gyotoku, 1926]. The inactivation is generally irreversible. The effect of quinine (free base only) is instantaneous, that of atoxyl (sodium *p*-amino-phenyl-arsenate) takes fifteen to thirty minutes to develop fully, and some protection is afforded by the substrate, which is not the case with quinine. The inhibition, over a wide range, is proportional, not to a power, but to the logarithm, of the concentration of poison. Exceptions are furnished by gastric lipase [Rona and Takata, 1923] which is reversibly poisoned by quinine and its derivatives, the inhibition varying as a power, not the logarithm, of the concentration, and by kidney lipase [Rona and Haas, 1923] where the relation between logarithm of inhibition and logarithm of atoxyl concentration is given by a curve analogous to a dissociation residue curve, and the inhibition is reversible. Different lipases differ enormously in their sensitivity. Table XIX. gives the approximate amounts of poison, in mgm. per litre needed for half inactivation.

TABLE XIX.

Lipase from—	Quinine.	Atoxyl.
Human serum .	2	1
Other serum .	300-1000	1
Liver . . .	200	·018
Pancreas . .	10	200
Stomach . .	100	?
Kidney . .	100	·07

There were no great differences between lipases from man and other mammals (dog, cat, horse, pig, rabbit, rat, guinea-pig, mouse) from any source except sera, which latter also differed amongst themselves, some being quite unaffected by quinine. In general when a sensitive and insensitive lipase were mixed, there was neither protection nor sensitization. Thus, in a mixture of equally active amounts of lipase from human serum, a purified and quinine resistant preparation from pig's pancreas, and pig's liver, the activity was reduced to $\frac{2}{3}$ by quinine (serum lipase poisoned) and to $\frac{1}{3}$ by atoxyl (serum and liver poisoned). There was only one exception. Serum protected liver (but not kidney) lipase from small amounts of atoxyl.

The interpretation of these results is somewhat controversial. Willstätter and Memmen [1924, 2] observed changes in sensitivity to quinine and atoxyl on purification of liver and pancreatic lipases, and found that the addition of albumin to a purified preparation of the latter increased the inhibition produced by quinine from 7 per cent. to 43 per cent. They regard the specific sensitivity as due to accompanying activators rather than to the enzymes themselves. Rona and Gyotoku [1926] purified a number of lipases by Willstätter's methods, and found that in general this caused a slight falling off in the sensitivity to poisons. On the other hand, the purification of rabbit's liver lipase rendered it insensitive to atoxyl, and that of pig's stomach rendered it sensitive to quinine. Even so mixtures did not exhibit mutual protection or sensitization. It is clear that if the specificity of the lipases is due to coenzymes, these must be very tightly bound to the corresponding enzymes.

A large amount of scattered work exists regarding the action of alkaloids and other drugs on various enzymes. While often inhibiting their action, they may occasionally accelerate it. For example, Rona and György [1920] found that atoxyl accelerates urease action.

A few attempts have been made to alter the composition of enzymes by substitution of well-known groups. Thus Helferich, Speidel, and Toeldte [1923] methylated, ethylated, and acetylated the β-glucosidase of emulsin with diazomethane, diazoethane, and acetyl chloride respectively. It was found that the degree of in activation ran parallel with the amount of methoxyl, ethoxyl, or acetyl groups introduced. Attempts to regenerate the enzyme were failures.

Proteases.

Apart from the effects described in the section on surface-active substances, the most important work has been done with other enzymes and with antibodies. So far as I know no enzyme has been destroyed by protease-free preparations of other enzymes; but a number of cases exist where proteases have been found to be effective, though it must be remembered that the protease preparations used generally, if not always, contained other enzymes. The most thorough investigation is of the hydrolysis of purified yeast saccharase by pepsin and trypsin (+ erepsin) [Euler and Josephson, 1924]. The peptic hydrolysis at pH 3·2 was followed by measurements of viscosity and enzymatic activity. Whereas the viscosity of a gelatin solution reached its minimum value (64 per cent. of the initial) in four hours, the viscosity of saccharase had only diminished by 2 per cent. in this time, and only by 3·4 per cent. after thirty-seven hours. In thirty-eight hours the enzymatic activity had only fallen by 14 per cent. The slowness of attack by pepsin was not due to inhibitors. Similarly, trypsin attacked saccharase far more slowly than gelatin or peptone at pH 6·0. The viscosity was hardly affected in forty-eight hours, though hydrolysis of gelatin and peptone was fairly complete in two. The enzymatic activity did not fall in the first twenty hours, but fell to

59 per cent. in forty-eight hours, to 28 per cent. in 120 hours. The protein or protein-like part of purified saccharase is therefore abnormally resistant to proteases. The falling off in enzymatic activity may either be due to the fact that the enzyme includes a protein-like portion, or that, as in the case of gastric lipase (see p. 175), it is protected from "spontaneous" inactivation by accompanying proteins.

Anti-Enzymes.

A number of anti-enzymes exist in nature. The best-known is the antitrypsin of normal serum. Hussey and Northrop [1923] showed that this substance behaves quantitatively like the trypsin inhibitor produced by the tryptic digestion of proteins, and hence the simplest hypothesis concerning its nature is that it is a polypeptide which unites with the enzymatically active group of trypsin.

Onslow [1915] found that the skins of newborn black, chocolate, and blue rabbits contain an enzyme which catalyses the production of a black pigment from tyrosine in presence of H_2O_2. This may be identical with a 3-4 dihydroxy-phenyl-alanine oxidase whose presence in such extracts has been shown by Haldane (unpublished). Onslow's enzyme was inhibited by extracts from two types of white rabbit skin due to dominant genes, namely the white parts of the "English" type of piebald and the white bellies of grey and yellow rabbits. For example, "English" skin extracts completely inhibited blackening by five times its volume of black skin extract. The inhibitor is destroyed by boiling, and disappears on standing for forty-eight hours at room temperature. It is precipitated by saturated, but not by half-saturated, ammonium sulphate.

The pancreas and other organs, and also *Aspergillus oryzae*, contain a heat-labile inhibitor of myozymase. Case and McCullagh [1928] showed that this had the same distribution as amylase, and like it, is a thermolabile colloid. Moreover, inhibitory and amylolytic powers of different preparations were roughly proportional. The inhibitor does not destroy any component of myozymase. The same amount of inhibitor caused 98 per cent. inhibition in a 2 per cent. starch solution, and only 18 per cent. with 6 per cent. starch. The inhibition is largely prevented by maltose, which is known to unite with amylase (see p. 63), but not by other sugars. The inhibitor prevents lactic acid formation from starch or glycogen, but not from glucose in presence of activator (see p. 144). The identity of the inhibitor with amylase is fairly clear. It does not act by destroying the starch, for inhibition may be complete even when the starch still gives a colour with iodine. The most obvious picture of the inhibition process is that the starch molecules may be so completely surrounded by amylase molecules as to prevent the access of other enzymes.

An amount of amylase which reduces the rate of lactic acid formation from a starch solution by 70 per cent., will destroy the iodine-blueing power of the same solution at 30° C. in about thirty minutes. Hence the time taken by a single amylase molecule to break down a starch molecule to the achromic stage is of the order of ten minutes. It cannot, at any rate, be of the order of ·001 second, as in the case of some other enzymes.

Dakin and Dudley [1913, 3] found that pancreatic extracts inhibit glyoxalase. The antiglyoxalase is a colloid and fairly thermolabile. It is not insulin, trypsin, or cozymase [Kuhn and Heckscher, 1926], nor does it inhibit aldehyde mutase, which therefore differs from glyoxalase. Pancreatic extracts give an orange precipitate with phenyl glyoxal [Foster, 1925]. Girsavicius (unpublished) has shown that this is due to a heat-stable base or group of bases, quite distinct from the true antiglyoxalase, which would seem to act on the enzyme rather than its substrate.

Another naturally occurring anti-enzyme is anticatalase, discovered by Battelli and Stern. A full review with references is given by Stern [1927]. It acts on mammalian catalase, and is found in a number of mammalian tissues. It may be freed from catalase and philocatalase by treatment with acids. It inhibits catalase, though never completely. The inhibition requires O_2 or a H acceptor, has an optimum pH of 6·4, and an optimum temperature of about 38°, the anti-catalase being rather thermolabile.

Other tissue extracts contain a thermolabile substance or group of substances, philocatalase. This may not only protect catalase from anticatalase, but regenerate the inactivated enzyme. Reducing substances such as aldehyde have a similar but weaker action. Boiled tissue juice increases the activity of philocatalase. It would seem that anticatalase is probably an oxidizing enzyme. The whole system seems worthy of much further study.

Much more work has been done on immune anti-enzymes, which appear when enzymes are injected into animals. Some of the older work is vitiated by lack of control of pH. An example of more critical work is the paper of Lüers and Albrecht [1926] on antiamylase. Partially purified but not protein-free malt amylase was used. Normal rabbit serum possessed no appreciable amylase or antiamylase action. After injection of amounts of amylase gradually increasing from 1 to 40 mgm. into rabbits (which were not made ill thereby) for periods of three to eight weeks the serum was found to contain antiamylase. Thus 2 c.c. of serum reduced the rate of hydrolysis of 100 c.c. 3 per cent. soluble starch solution by 5 mg. of amylase at pH 5 and 37° C. by 38 per cent. to 60 per cent. The inhibition was increased, as is usual with immune sera, if the enzyme and serum were previously incubated. It was proportional to the amount of serum up to about 80 per cent. inhibition, but even with large amounts of serum complete inhibition was not obtained. A precipitin reaction appeared with ·001 per cent. amylase solution, but not with ·0001 per cent., which was, however, inhibited. The inhibition did not extend to pancreatic or salivary amylase, nor was malt amylase affected by precipitin formation in an ovalbumin-antiovalbumin system. Only very slight amounts of antiamylase were obtained on injecting amylase inactivated by moderate heat, although the precipitin reaction (presumably due to other antigens) remained. It is clear that in this case it cannot be decided whether the immunity was developed to the enzyme itself, or to colloids closely associated with it. The same doubt applies to all the very numerous cases in which immune anti-enzymes have been obtained. Even were immunity obtained with a highly purified enzyme, it would throw little light on its nature, since carbohydrates and lipoids, as well as proteins, can unite with immune bodies.

The work of Bach and his colleagues renders it probable that in many cases the immunity is due to accompanying substances rather than the enzyme itself. Engelhardt [1924] had shown that alumina, ferric hydroxide, kaolin, or charcoal, after treatment with antiphenolase serum, adsorbed the enzyme. Bach, Engelhardt, and Samysslov [1925] applied this technique to rabbit antisaccharase. They found that whereas the antiserum alone reduced the activity of crude yeast saccharase by 10 per cent. to 15 per cent., it was reduced by 80 per cent. to 100 per cent. in presence of kaolin. The adsorbed enzyme could, at least in part, be eluted. When the experiment was repeated with a purified saccharase, only 30 per cent. to 40 per cent. inhibition was obtained with serum and kaolin together. But fairly strong inhibition was obtained after immunizing with heat-inactivated saccharase. The existence of specific antisaccharase is thus rendered very doubtful.

CHAPTER IX.

THE PURIFICATION AND CHEMICAL NATURE OF ENZYMES.

THE purification of enzymes is a matter of extreme difficulty. The objects to be gained are several. It may be desired to obtain insight into the chemical nature of an enzyme, to separate it from other enzymes, or from accelerating or inhibiting substances; or merely to show that it is not, e.g. a protein, or does not contain phosphorus. According to the goal desired the procedure will be different. Thus the adsorption method has not led to the production of very active samples of trypsin, but it has allowed the separation of this enzyme from erepsin, kinase, and so on (see p. 109). Again by substituting arsenate for phosphate at a certain stage in the preparation of yeast saccharase, it was possible to show that saccharase contains extremely small amounts of phosphorus [Euler and Josephson, 1923, 1].

As regards the degree of purification achieved, Table XX. summarizes the more successful attempts. The number given is the ratio

TABLE XX.

CONCENTRATION OF ENZYMES.

Pancreatic lipase	240	Willstätter and Waldschmidt-Leitz (1922).
Gastric lipase	600	Willstätter and Bamann (1928).
Liver lipase	128	Kraut and Rubenbauer (1928).
Ricinus lipase	100	Willstätter and Waldschmidt-Leitz (1924).
Almond β-glucosidase	60	Josephson (1926).
Yeast saccharase	1,200	Willstätter and Schneider (1925).
Pancreatic amylase	125	Willstätter, Waldschmidt-Leitz, and Hesse (1924).
Aspergillus amylase [1]	30	Sherman and Tanberg (1916).
Rennin	55	Fenger (1923).
Jack Bean urease [2]	730	Sumner and Hand (1928).
Liver catalase	220	Euler and Josephson (1927, 1).
Milk xanthine oxidase	500	Dixon and Kodama (1926).
Horse-radish peroxidase	20,000	Willstätter and Pollinger (1923).
Myozymase [3]	50	Meyer (1928).
Lactarius quinol oxidase [4]	630	Wieland and Sutter (1928).

[1] Compared with commercial preparations. Really > 100.

[2] As the original material was defatted the total concentration achieved was greater.

[3] On aqueous extract.

[4] ? On wet weight of fungus.

of enzymic activity per gram dry weight of final preparation to the same per gram dry weight of original material. In some cases the former was a whole organ, in one or two an extract. It must be emphasized that in many cases the dry weight of the enzyme preparation was only obtained after dialysing away other substances (e.g. glycerol in the case of lipases), and the dry preparations so obtained were often partly inactivated. For this reason Sherman, Caldwell, and Adams [1926] have criticized the method in question, holding that part of the preparations dialysed out. But such criticism does not hold concerning other preparations, such as peroxidase. Again in some cases a falsely high yield has been obtained by the removal of inhibitory substances. However, in the case of none of the enzymes which have been highly purified has an increase of as much as 100 per cent. in the total amount of enzyme been observed at any stage. So the figures given for concentration achieved are substantially correct.

Enzyme Units.

The unit of any enzyme is the amount which will do a certain amount of work under well-defined conditions. Thus Willstätter has employed, *inter alia*, the following units :—

Saccharase Unit (S.E.). The amount of enzyme which will reduce the rotation of 4·0 gm. cane sugar in 25 c.c. of 1 per cent. NaH_2PO_4 solution at 15·5° C. to 0° (after adding alkali to cause mutarotation of α-glucose) in one minute. This corresponds to 75·5 per cent. of hydrolysis. (The pH is optimal.) [Willstätter and Kuhn, 1923.]

Peroxidase Unit (P.E.). $\frac{1000}{x}$ times the amount of enzyme needed to produce x mg. of purpurogallin in a solution of 5 gm. pyrogallol + 50 mg. H_2O_2, dissolved in 2 litres of H_2O at 20·0° in five minutes; where x must lie between 15 and 25 [Willstätter and Pollinger, 1923].

Lipase Unit (L.E.). The amount of enzyme which, in a solution of 2 c.c. N NH_3 + NH_4Cl buffer of pH 8·9, 10 mg. $CaCl_2$, and 15 mg. egg albumin, with the addition of 2·5 gm. olive oil, of saponification value 185·5, all in a volume of 13 c.c. at 30°, causes 24 per cent. hydrolysis in one hour [Willstätter, Waldschmidt-Leitz, and Memmen, 1923].

Butyrase Unit (B.E.). The amount of enzyme which, in 56 c.c. saturated tributyrin solution, 2 c.c. of N NH_3 + NH_4Cl buffer of pH 8·6, 10 mg. $CaCl_2$, 10 mg. Na oleate, and 30 mg. egg albumin, causes the number of drops per minute from an Ostwald's stalagmometer giving one drop per second with pure water, to diminish by twenty at the end of fifty minutes at 20° C. [Willstätter and Memmen, 1923].

In the case of saccharase the amount of substrate transformed is proportional over wide ranges both to the enzyme concentration and the time. In that of peroxidase this is not the case, and the conditions must be exactly as given. In the case of lipase, activators are used, since in the course of purification substances which can act as coenzymes are removed. Two distinct units are employed, and whereas in the case of pancreatic lipase 1 L.E. = about 1000 B.E., in that of liver esterase 1 L.E. = 10 B.E.

The purity of a preparation is generally measured by the number of units per gm. (or mg.) of dry weight, but for the reasons above given Sherman prefers to use the number per mg. N. This is, of course, inapplicable where carbohydrate or fat are likely to occur as impurities.

The Extraction of Enzymes.

Some enzymes are very readily extracted with water from tissues, others with greater difficulty. Thus muscle succinoxidase is not extracted by repeated washing of chopped muscle with water, whence Battelli and Stern [1922] at first regarded it as a complex and insoluble ferment or oxydone; but then found that it could be extracted in mildly alkaline water, best with Na_2HPO_4 solution [Ohlsson, 1921], though it is rather easily removed from solution. The solubility of the same enzyme may differ in different species. Thus histozym can easily be dissolved from the kidney of the dog, partially from that of the pig, and not at all from that of the horse [Waldschmidt-Leitz, 1926]. Pancreatic lipase, though soluble in water, is unstable, and it is more satisfactory to dissolve it in glycerol [Willstätter and Waldschmidt-Leitz, 1923]. However, *Ricinus* lipase is apparently insoluble either in water or glycerol [Willstätter and Waldschmidt-Leitz, 1924], and is, moreover, destroyed by these liquids.

The enzyme may often be removed from insoluble cell constituents by altering the pH, using a non-aqueous solvent such as glycerol, and so on. Sometimes it is found desirable to allow the cells to autolyse, the enzyme being tightly bound to the structure of the intact cell. This is the usual procedure in the preparation of yeast enzymes. However, yeast maltase is very sensitive to acids, and is destroyed by the acids produced by autolysis in unbuffered solutions. In some cases, however, the concentration of the enzyme is facilitated, if it is not removed too soon from insoluble substances. Thus Willstätter and Stoll [1918], and Willstätter and Pollinger [1923], found that the most satisfactory peroxidase preparations were obtained from horse-radish by allowing finely divided plant material to dialyse first in presence of water and then of oxalic acid. A fraction of the enzyme is lost, but also very large amounts of substances of low molecular weight. The enzyme required from the preparation is only liberated on neutralizing the oxalic acid. In this way the enzyme can be concentrated several hundred times. It is often desirable to remove lipoids with suitable solvents before the extraction of an enzyme is begun.

Methods Depending on Solubility.

A considerable degree of purification can be obtained by methods involving the solubility of the enzyme in question. In particular it can be separated from other colloids by precipitation with alcohol at a suitable pH. Many commercial preparations are made in this manner. As an example of this type of purification Willstätter and Csanyi's [1921] method for emulsin may be described. The seeds were dried, ground and most of the oil pressed out. The remainder was extracted with ether, and dried, losing in all 66 per cent. of its weight, but very little enzyme. The powder was extracted with N/10 NH_3, which dissolves more than half of it. On neutralizing with acetic acid most of the solids, but little enzyme, are precipitated. The remaining fluid is precipitated with four times its volume of alcohol. This precipitate is washed with alcohol and ether, and dried. Some of the proteins are thus rendered insoluble, and the emulsin can be separated from them with water.

Sherman and Schlesinger achieved a considerable degree of purification of malt and pancreatic amylases by this method. In general it seems to be less effective than adsorption. However, Sumner [1926, 1, 2] has prepared an exceedingly active urease preparation by a very simple method. Defatted Jack Bean meal is stirred up with 31·6 per cent. (by volume) acetone solution, and the mixture filtered through a Schleicher and Schüll filter. The filtrate is placed in the ice chest, when a protein crystallizes out. The crystals are washed with more acetone solution and drained. They can be dissolved in distilled water and impurities centrifuged off. Recrystallization is possible, but involves considerable loss. Only very active preparations of Jack Bean meal behave in this way. When these are not used, it is necessary to add dilute acetic acid in order to bring the urease out of solution. The enzyme is then contaminated with other substances, and only about half as active as Sumner's best preparations [Sumner and Hand, 1928]. Sumner believes that the crystalline protein is the enzyme. It is soluble in water, coagulated by heat, and gives all the protein and no carbohydrate tests. On recrystallization the activity per mg. N is unaltered. The purified enzyme is extremely susceptible to traces of metal, and glass-distilled water must be used.[1]

Enzymes can often be regained after precipitation by such reagents as tannin. This has been used in the purification of peroxidase by Willstätter and Pollinger [1923], who then acidified the precipitate and removed the enzyme from the resulting solution by adsorption. Safranine has been used by Marston [1923] and Forbes [1927] in the purification of proteases which it precipitates. Forbes was able to purify pepsin about twenty-fold with its aid. The purified enzyme had the general characteristics of a very acid protein, and contained little hexone bases, but large amounts of mono-amino-acids.

Purification by Adsorption.

This method, which was developed by Willstätter and his colleagues, but has also been applied by Euler and his colleagues, by Dixon and Kodama, and others, appears to be the most widely applicable method of purifying enzymes, and in particular of separating one enzyme from another. It is based on the fact that a variety of adsorbents remove enzymes from solution, and that the enzyme can generally, though not always, be eluted from the adsorbent by altering the pH and thus the charge on the enzyme, adsorbent, or both, or by adding a substance which is adsorbed in preference to the enzyme. Kraut [1928], who gives a full account of this method, gives a list of forty substances which have been used to adsorb enzymes. The most important of these are a variety of aluminium hydroxides, kaolin (aluminium silicate), and charcoal.

Michaelis and Ehrenreich [1908], who rediscovered the method of separating enzymes by adsorption, believed that this process depended on the charge of the enzyme and adsorbent. Thus, kaolin is negatively charged and adsorbs basic dyes, alumina positively and adsorbs acid dyes. Substances adsorbed by both are regarded as ampholytes. In alcoholic solution an ampholyte generally develops acid properties, and may be adsorbed on a basic substance such as alumina which

[1] Northrop [1930] has since isolated pepsin as a crystalline protein.

will not remove it from aqueous solution. Now the adsorption of an enzyme may depend on the presence along with it of other substances. Thus Willstätter [1922] found that yeast saccharase was not adsorbed by kaolin from the crude solutions, but was so when purified, while alumina adsorbed amylase from pancreatic extracts, but not from purer solutions. It is clear then, that the adsorbability of an enzyme may depend, not on its own charge, but on that of a complex formed by it with some other substance. Again, small differences in the adsorbent, which do not alter its charge, may affect its affinity for different enzymes in opposite directions. Thus Willstätter and Kraut [1923, 1924] describe at least six different aluminium hydroxides, two of which absorb yeast maltase preferentially to saccharase, the others having the opposite effect.

Other things being equal, however, it is undoubtedly true that pH is a determining factor in adsorption. Numerous examples are given by Kraut [1928]. At a given pH the adsorption isotherm for a given enzyme and adsorbent may have various forms. In the "normal" type a maximum amount of enzyme is adsorbed at a certain concentration, and this remains constant when more enzyme is added. If substances are present which are adsorbed, and with which the adsorbent is saturated more readily than with the enzyme, very little enzyme is adsorbed at first, and the isotherm is concave upwards. In the opposite case a maximum of adsorption is reached at a certain concentration, and the amount then diminishes. In this case the adsorption is better from dilute than from concentrated solutions. When the adsorption isotherm is not "normal" the enzyme may be purified from its co-adsorbed concomitants by fractional adsorption. With a normal isotherm this is impossible. Adsorption may be used to remove impurities. Thus Dixon and Kodama [1926] found that they could remove 80 per cent. of the protein from their xanthine oxidase solution with charcoal, but that after this enzyme and protein were adsorbed in about equal amounts.

Elution may be carried out by changing the pH or by adding a specific eluent. Thus peroxidase, a basic substance, is adsorbed by kaolin from a solution in dilute acetic acid, a number of glucosides being left behind. It is eluted with dilute NH_3 [Willstätter and Pollinger, 1923]. On the other hand, saccharase can best be removed from alumina with a trivalent anion such as phosphate or arsenate. Fractional elution is often possible.

Other Methods.

Dialysis and electrodialysis have been employed at various stages. These may serve to remove impurities, and also to cause precipitation either of the enzyme or accompanying substances. Usually membranes impermeable to the enzyme were employed, but Nelson and Morgan [1923] were able to separate yeast saccharase from coloured impurities by membranes permeable only to the enzyme. Other enzymes have been used to destroy impurities. For example, Willstätter and Bamann [1928] employed proteolytic enzymes from yeast and other sources to destroy proteins and peptides which accompany gastric lipase. Dixon and Kodama [1926] used rennin to coagulate the caseinogen from which their xanthine-oxidase was otherwise inseparable. Heat and poisons have been employed to inactivate one enzyme in a mixture, and so on. Finally, considerably increased purity may sometimes be obtained by biological methods. Thus Willstätter, Lowry, and Schneider [1925] found that when yeast was placed in very dilute cane sugar or glucose solutions, and allowed to ferment for some hours, its saccharase content was increased ten to fifteen times. From this Willstätter, Schneider, and Wenzel [1926] prepared the most active preparation of saccharase so far secured (1 gm. = 238 units), but its activity was only about double that obtained from ordinary yeast.

Some examples of purification are given below. Fuller accounts are given by Waldschmidt-Leitz [1926], Kraut [1928], and Grassmann [1928].

Concentration of Peroxidase.

The following example is taken from Willstätter and Pollinger [1923], and Willstätter and Stoll [1918] :—

1. The horse-radish is cut up coarsely and allowed to soak in water.

2. After washing it is soaked in dilute oxalic acid and washed.

3. Dilute baryta is cautiously added, and the liquid thrown off until it becomes alkaline. The fractions which are alkaline contain the enzyme.

4. The baryta is precipitated with CO_2, and small amounts of alcohol added to precipitate mucin-like substances.

5. Impurities are adsorbed by alumina in aqueous solution, the enzyme being unaffected.

6. A series of adsorptions were carried out from 50 per cent. alcohol. It was twice adsorbed on alumina, the residual liquid

decanted, and elution carried out with water saturated with CO_2. It was then adsorbed with kaolin, and eluted with ammonia, which was subsequently neutralized. This removed carbohydrates, but some kaolin went into colloidal solution. It was then adsorbed by alumina and once more eluted. A fourth adsorption with alumina caused a loss of enzyme, a fifth a still greater, but the solution (about ·013 per cent.) had the highest peroxidase activity ever recorded, after concentration and standing for forty hours on ice.

Concentration of Saccharase.

Numerous methods are given by Schneider [1928]. The following gives powerful and stable preparations with ease, but other more complicated methods have yielded substances about twice as active.

Yeast grown in dilute sugar solution is warmed to 30° C. It is kneaded with 10 per cent. of its weight of toluene, and liquefies in about thirty minutes. After another one to three hours it is diluted with two to three volumes of water and centrifuged. 200 c.c. of the supernatant autolysate are mixed with 1100 c.c. of water and 250 c.c. of 2 N acetic acid. 21·3 gm. of kaolin previously treated with HCl are added, and filtered off by suction. After washing, the kaolin is suspended in distilled water, and normal soda added till the reaction is just alkaline. The kaolin is mostly centrifuged off, the remainder being removed with kieselguhr on a suction pump. The resulting solution is dialysed. The yield may be as high as 77 per cent. of the saccharase in the autolysate.

Concentration of Xanthine-Oxidase.

Dixon and Kodama [1926] mixed milk at 35° with rennin, allowed it to clot, broke up the curd, and centrifuged. The fluid was mixed with an equal volume of saturated ammonium sulphate. The precipitate (which rose to the surface) was dried between filter papers, extracted with ether, squeezed again between filter papers, and dried *in vacuo* to avoid oxidation. 5 gm. of this preparation were dissolved in 250 c.c. H_2O, shaken with 5 gm. of charcoal, and filtered. This was done five times, and removed most of the protein, but little of the enzyme. The filtrate was brought to pH 5, and adsorbed on 10 gm. kaolin. After filtering by suction, 100 c.c. 1 per cent. Na_2CO_3 were added to elute. To the resulting solution an equal volume of saturated ammonium sulphate solution was added, and the precipitate, after squeezing, dried *in vacuo*. About 50 per cent. of the total enzyme of the milk was thus obtained.

Chemical Nature of Purified Enzymes.

The only preparations which have any possible claims to approximate purity are those of lipase, saccharase, urease, xanthine-oxidase, and peroxidase of Table XX. Urease has the properties of a protein. Sumner describes it as a globulin, but it is soluble in distilled water. Dixon and Kodama's [1926] xanthine-oxidase showed certain protein reactions. Willstätter and Pollinger's peroxidase contained (ash free) 46·0 per cent. to 49·4 per cent. C, 6·9 per cent. to 8·6 per cent. H, 9·4 per cent. to 13·6 per cent. N in different preparations. P was only present in traces, and only about 0·1 per cent. Fe was present in some of the purest preparations. Up to a certain point there was a marked proportionality between Fe content and peroxidase activity. But when the enzyme was purified by precipitation with tannin, which gave very active preparations, the amount of Fe could be reduced from ·11 per cent to ·06 per cent., while the activity was increased by 50 per cent. Preparations were obtained free from carbohydrate, Millon, and ninhydrin reactions. A good sample was precipitated by tannin and KI_3 but not by $HgCl_2$ or uranyl acetate, and only gave a faint cloudiness with phosphotungstic acid. On heating it charred and gave a smell of roasted coffee, but no pyrrole reaction. The most striking property of the preparations is their reddish-brown colour, resembling that of pyrroporphyrin, and the authors consider it probable that the enzyme is coloured. On the other hand, it is certain that, if so, it is accompanied by other coloured substances, and that inactivation does not necessarily lead to loss of colour. In view of Keilin's [1925] tentative identification of the peroxidase of animals (though not of higher plants) with cytochrome or one of its components, the hypothesis that peroxidase is a porphyrin derivative gains in plausibility.

Preparations of gastric, pancreatic, and hepatic esterases, and of yeast saccharase and pancreatic amylase have been obtained free from protein reactions, in particular the biuret, Millon, ninhydrin, and tryptophan reactions [Willstätter and Bamann, 1928]. The amount of protein in these preparations must certainly be small. If, as many workers believe, the enzymes are all proteins, it is certainly remarkable that the majority of the successful attempts to purify them have led to the obtaining of substances which are at least predominantly non-proteins, although the original material from which they were derived consisted largely of protein. It should be remarked that the purest preparations, which give no protein reactions, are still non-dialysable and, where analysed, contain C, H, O, and N.

However, the proteins or protein breakdown products which are associated with enzymes are of the greatest significance for their stability. Thus Willstätter and Bamann [1928] found that, as soon as the last substances giving protein reactions are removed from gastric lipase (by which its activity is about doubled), its stability is very greatly diminished. Kraut and Rubenbauer [1928] found the same with hepatic esterase. As soon as a substance giving Millon's reaction was removed the stability fell suddenly. Whereas impure preparations lost 10 per cent. to 25 per cent. of their activity in the course of several days, and then became more stable, the purified enzyme lost 50 per cent. of its activity over night, and became wholly inactive on further standing. If, then, the protein is not part of the catalyst, it is a part of the enzyme considered as a stable substance. Willstätter, Graser, and Kuhn [1922] point out that while such properties of yeast saccharase as pH optimum, activity, and affinities are hardly affected by purification, its heat inactivation and optimum temperature are greatly altered. Thus an autolysate suffered no inactivation at 52°, while a purified preparation lost 43 per cent. of its activity. In the case of peroxidase the instability of the purified enzyme may show itself in increased activity. Thus Willstätter and Weber [1926, 1] found that while horse-radish juice showed changes in its peroxidase activity not exceeding + 3 per cent. and — 2 per cent. in 120 hours, a pure preparation increased its activity by 22 per cent. in two hours, fell to 74 per cent. of the original value on the second day, rose to 89 per cent. on the third, fell to 51 per cent. on the fourth, and had risen to 61 per cent. on the sixth. Estimations of activity with pyrogallol and leuco-malachite green gave proportional values at any given time.

Willstätter and his colleagues (especially Schneider and Wentzel, 1926) devoted much labour to obtaining highly active saccharase preparations free from various reactions. Thus in the products of relatively slow autolysis the enzyme is accompanied by a tyrosine-containing substance, and can be obtained tryptophan-free. When autolysis is rapid it is accompanied by a tryptophan-containing substance, and good preparations may contain up to 9 per cent. of tryptophan. This can be reduced by the addition of peptides such as leucyl-glycine, leucyl-glycyl-glycine, and glycyl-tyrosine. The tryptophan per unit of enzymic activity can then be reduced by adsorption. Euler and Josephson [1924] found that pepsin and trypsin attacked yeast saccharase very slowly (see p. 162). The authors regard these results as demonstrating that saccharase is a specially stable protein. But they can also be explained as due to the destruction of a protective colloid. In view of Sumner and Hand's [1928] important observation that crystalline urease preparations can be protected by gum-arabic, egg albumin, and H_2S from traces of metallic impurities (probably lead) found in ordinary distilled water which otherwise halve its activity within a minute, it would seem possible that, with suitable precautions against metallic poisoning, protein-free enzyme preparations of relative stability might be obtained.

Sumner's urease preparations consist of a crystalline protein, and he regards them as pure urease. Here, too, it is conceivable that the protein is acting both as a protective colloid and a specific adsorbent, though he finds that an associated protein, concanavalin B, has very slight adsorptive properties. On the other hand, the high activity found is no conclusive argument for purity. His preparation is 730 times as active as the original meal, as compared to a figure of 12,000 to 20,000 in the case of peroxidase, and hydrolyses 0·2 of its weight of urea per second at 20°, as compared with an activation of 264 times its weight of H_2O_2 per second in the case of peroxidase. It is therefore conceivable, though not necessarily probable, that it contains only a small fraction of actual enzyme.

While Euler regards enzymes as definite substances, Willstätter [1922] and Fodor [1926] believe them to consist of an active group and a colloidal carrier. The contradiction is not fundamental if it be admitted that the colloidal carrier is necessary to the stability of the enzyme.

Some Applications of Molecular Statistics.

Willstätter and Pollinger's best peroxidase could activate 1000 times its weight of H_2O_2 per second at 20° when used on leuco-malachite green, which is oxidized 120 times as fast as pyrogallol. The H_2O_2 concentration was 2·5 mg. per litre. At 1 mg. per litre the oxidation proceeds 2·6 times as fast [Willstätter and Weber, 1926, 1]. We will assume, that as with inorganic catalysts, activation can only occur when a substrate molecule collides with an active centre on the surface of the enzyme. In a gas the mass striking an area

A per second is $M = \frac{3}{13}\bar{u}\rho A$, where $\bar{u}$ is the root mean square velocity of the gas molecule, ρ its density [Hinshelwood, 1926]. It is believed that similar figures, with a small correction, which may be neglected in rough calculations, apply to aqueous solutions. The enzyme being a colloid its velocity may be neglected. For unhydrated H_2O_2 $\bar{u} = 4{\cdot}3 \times 10^4$ cm./sec. at 20°, while even if the molecule were hexahydrated this value would only be halved,

$$\rho = 10^{-6}, \therefore M = \frac{3}{13} \times 4{\cdot}3 \times 10^4 \times 10^{-6}A = 10^{-2}A,$$

or slightly less. Now the area of an H_2O_2 molecule is probably about that of an O_2 molecule, i.e. between 9·1 and $5{\cdot}8 \times 10^{-16}$ cm.^{2} It is probable that the active centre is of about the same dimension, so that for activation the centre of an H_2O_2 molecule must strike within an area of not more than 4×10^{-15} cm. Hence the mass of H_2O_2 colliding with an active centre per second is 4×10^{-17} gm. Now, if the molecular weight of enzyme per active centre be m, the weight colliding per chemical unit (weight of H atom) is $\frac{4}{m} \times 10^{-17}$, and per gram $\frac{4}{m} \times 10^{-17} \times 6{\cdot}06 \times 10^{23}$, i.e. $\frac{2{\cdot}42}{m} \times 10^7$ gm., or less. Thus, if $m = 24{,}000$ the amount colliding per gram per second would be 1000 gm. Supposing the enzyme to consist of an aggregate of molecules of the size of hæmatoporphyrin (molecular weight, 590), each possessing one active centre, the amount would be 40,000 gm. Finally, supposing the enzyme to be a sphere of radius 7×10^{-8} cm., corresponding to a molecular weight of about 1000, and all its surface active, the area would be $6{\cdot}2 \times 10^{-14}$ cm.^{2}, the mass colliding per second $6{\cdot}2 \times 10^{-16}$ gm., and the mass colliding per gram 380,000 gm. The mass activated per gram of purified peroxidase is, however, about 2600, hence 0·7 per cent. to 100 per cent. of the colliding molecules must be activated.

This calculation involves the assumption that the purest preparations would behave with regard to the two substrates quantitatively as do less pure preparations. Now, the temperature coefficient of peroxidase acting on leuco-malachite green is $Q_{10} = 2{\cdot}0$ over a range from 0° to 25° [Willstätter and Weber, 1926, 1]. The concentration is not given, but is very probably about 25 mg. H_2O_2 per litre. It is possible that at sufficiently low H_2O_2 concentrations there is no temperature coefficient. Otherwise there must be some very serious error in the assumptions underlying the above calculation, even if the number of colliding molecules is too great by several powers of 10. If, however, the above calculation is anywhere near correct, it is clear that peroxidase cannot be purified very much further.

The Number of Enzyme Molecules per Cell.

One gram of brewer's yeast contains about ·015 S.E. [Schneider, 1928]. A yeast cell with diameter 7 μ and density 1·1 weighs 2×10^{-10} gm., and hence contains 3×10^{-12} S.E. Now, 1 S.E. of the purest saccharase so far prepared (S.W. 11·9) weighs $\frac{50}{11 \cdot 9}$ or 4·2 mg. Hence yeast contains 63 mg. of this substance per kilogram wet weight, or about ·024 per cent. of its dry weight. A cell contains $1 \cdot 26 \times 10^{-14}$ gm., the weight of $7 \cdot 6 \times 10^{9}$ H atoms. Saccharase has a molecular weight about 50,000 (see p. 179), so the number of molecules per cell is less than 150,000. How much less depends on the impurity of the purest preparations. Thus, if they contain 10 per cent. of enzyme, the number is about 15,000. In yeast grown in weak sugar solution the figure would be ten or fifteen times as high.

Yeast is a highly specialized organism, very rich in saccharase. *Aspergillus niger* has only 2 per cent. to 3 per cent. as great an activity, and most phanerogam tissues far less. If their saccharases have about the same activity per gram as that of yeast, the number of molecules per cell must be less than 1000. Modern genetics suggest that genes are catalysts, the number per cell being 2, 1, or 0. It is probable that the number of molecules of a rare enzyme per cell may be of the same order.

CHAPTER X.

THEORIES OF ENZYME ACTION, AND CLASSIFICATION OF ENZYMES.

Some Chemical and Physical Properties of Enzymes.

THE attempts to purify enzymes have led to no definite conclusion as to their chemical nature. Except Sumner's urease, none of the most highly purified preparations appeared to be proteins, though all so far analysed contain C, H, O, and N. Experiments on irreversible inactivation by heat, nitrous acid, and so forth, point to a protein-like nature, but in view of the great instability of certain deproteinized enzymes they can be interpreted as due to the destruction of protective substances rather than the enzymes themselves. The evidence from reversible inactivation is more satisfactory. As noted in Chapter IX. it points to the possession by enzymes of acid, amino, and aldehyde groups, and in some cases, atoms of a heavy metal. Iron is frequently found in purified enzyme preparations, e.g. peroxidase. Willstätter and Pollinger's [1923] data show that peroxidase as active as their best preparation need not contain more than 0·10 per cent. of iron. Hence if the theory of p. 177 is correct, iron cannot be part of the active centre, for if so the molecular weight per active centre would exceed 55,000.

Enzymes usually behave as colloids, but their colloidal properties may become less apparent on purification, or may alter. In the case of yeast saccharase Euler and Ericson [1922] using a method depending on diffusion, found a molecular weight of about 20,000. However, the same method applied to egg albumin gives a weight of only 14,000, which is certainly too low. The molecule of saccharase is, therefore, probably larger than that of egg albumin, the true weight being somewhere about 50,000. Few enzymes will pass through collodion membranes of the type which excludes the average protein. Trypsinogen, however, passes through ultra-filters of such hardness as to hold back proteins [Bechhold and Keiner, 1927]. They regard trypsinogen as a semicolloid, or perhaps, even a crystalloid. On the other hand, kinase is considered to be a larger molecule than albumin or hæmoglobin. Olitzky and Boez [1927] find that peroxidase (of unstated origin, and perhaps heat-stable) passes through an ultra-filter which retains blue litmus, and is therefore of the dimensions of a large crystalloid.

Warburg has developed a highly ingenious method for the physical investigation of a particular catalyst (probably to be regarded as an enzyme) *in vivo*. The oxygen activator (oxygenase, Atmungsferment) of living cells is competitively inhibited by CO. The CO compound, like CO-hæmoglobin, is sensitive to light, and in consequence when, in a given mixture of O_2 and CO, the rate of O_2 uptake of a group of cells has been reduced to a fraction of that in absence of CO, it can be brought back towards normal by radiation. Comparing the efficiency of light of different frequencies, Warburg and Negelein [1929, 1, 2] have obtained a very satisfactory absorption spectrum of the CO compound. It has two bands at 430 $\mu\mu$ (blue), and 285 $\mu\mu$ (ultra-violet), and a general absorption over a wide range. The spectrum resembles those of CO-chlorocruorin and CO-protohæm (CO-reduced hæmatin), and renders it probable that the Atmungsferment is a compound of similar nature. It must, however, be remembered that Keilin [1929] discovered an oxygenase whose CO compound is not light sensitive.

Activation of the Substrate.

It is clear from the evidence of former chapters that the substrate unites with the enzyme, but does not in general cover the whole surface of the latter. The union depends on the structure of the substrate molecule, since molecules of related type unite in a similar way. Further, in many, if not all, cases, this union may be regarded as a chemical combination obeying the mass-action laws. Since, in principle, all enzyme action is reversible, every enzyme must, on thermodynamical grounds, be capable of union with at least one of the products of reaction. In many cases one or more of these products are highly labile molecules such as (2, 5) fructose, or the precursor of maltose in amylolysis. But in others, notably prunase, the enzyme has been shown to have a definite affinity for one or both of the immediate products of the reaction. It is interesting that no case is known suggesting that the enzyme-substrate union is bimolecular except where two different molecules are synthesized (e.g. lipase). For example, catalase obeys the Michaelis-Menten equation with great accuracy. It is therefore probable that the reaction catalysed is $H_2O_2 \rightarrow H_2O + O$, followed by the uncatalysed reaction $2O \rightarrow O_2$.

Modern theories of chemical reaction are based upon the idea of an active molecule susceptible of spontaneous chemical change. Those members of a molecular species which possess more than a certain amount of energy (whether in the molecule as a whole or in certain of its degrees of freedom) are active. The amount of energy can, in certain cases, be calculated from the temperature coefficient of the reaction. It is always smaller for catalysed reactions than for the same reactions when uncatalysed, both in the cases of non-enzymatic and enzymatic catalysis. Thus the enzyme-substrate compound requires less energy for its activation than the substrate alone. The difference between the energies of activation of two antagonistic reactions is, of course, equal to the free energy of the reaction.

If E be the energy of activation, and Q_{10} the temperature coefficient at a mean temperature T, then $\log_e Q_{10} = \frac{10E}{RT^2}$, approximately. A rise of 10° in T in the neighbourhood of T = 300 (27° C.) should, therefore, only reduce Q_{10} in the ratio $Q_{10}^{\frac{-1}{15}}$, or if $Q_{10} = 2{\cdot}0$ at 27° C., it should be reduced to 1·91 at 37°. A glance at Table XII. makes it clear that the observed values of Q_{10} for almost all hydrolytic enzyme actions fall off very much more rapidly than this with temperature, and hence the Arrhenius equation connecting reaction velocity with energy of activation is not obeyed. The calculated values of the energy of activation have, therefore, little significance. This is true even in the case of yeast saccharase, where, as pointed out in Chapter IV., fallacies inherent in the calculation in the case of most enzymes can be avoided. A possible explanation of the rapid diminution with temperature of the apparent energy of activation may be as follows.

Calling the enzyme and substrate molecules E and S, the molecule ES requires a certain energy of activation A before it will undergo chemical change. Now, a newly-formed ES molecule will possess more energy than the average, because it has been formed from activated molecules, which, moreover, may have gained energy by combination if the equilibrium $E + S \rightleftharpoons ES$ has a temperature coefficient. Hence the ES molecules which take part in the catalysed reaction will fall into two groups: (*a*) Those which react immediately after combination. These are already possessed of some extra energy, and will therefore require an additional energy increment less than A. (*b*) Those which react after existing for a period during which they have lost their initial extra energy. These will require the full energy increment A. Hence the actual temperature coefficient of the velocity may be expected to be a compromise between those characterizing molecules of types (*a*) and (*b*). However, further *ad hoc* assumptions are needed to explain the rapid fall with temperature in the apparent critical increment.

Speculation has been largely directed to the nature of the activation process. Let us consider the case of the hydrolysis or synthesis of salicin by β-glucosidase. The synthesis is, of course, a very slow reaction, but must be assumed to occur. Both β-glucose and saligenin combine with the enzyme, each uniting at a different spot. Josephson's arguments (p. 49) go far to show that the glucose and saligenin residues of salicin combine respectively at these same spots. They must, therefore, be near together on the enzyme molecule. When a glucose and saligenin molecule are both so combined with the enzyme they require less than the usual amount of energy increment to make them unite with loss of a water molecule, and when a saligenin (or saligenin-hydrate) molecule is so combined it requires less than the usual energy increment to undergo hydrolysis. One way in which this condition might be fulfilled would be if the molecules when combined with the enzyme, lay slightly further apart than their equilibrium distance when combined as salicin, but nearer than their equilibrium distance when free. The enzyme would thus tend to pull the salicin molecule apart, but to push the molecules of glucose and saligenin together (Fig. 35). In view of the very wide specificity of β-glucosidase, such a theory is, however, somewhat doubtful. The same hypothesis might clearly be applied to other hydrolytic enzymes. Using Fischer's lock and key simile, the key does not fit the lock quite perfectly but exercises a certain strain on it.

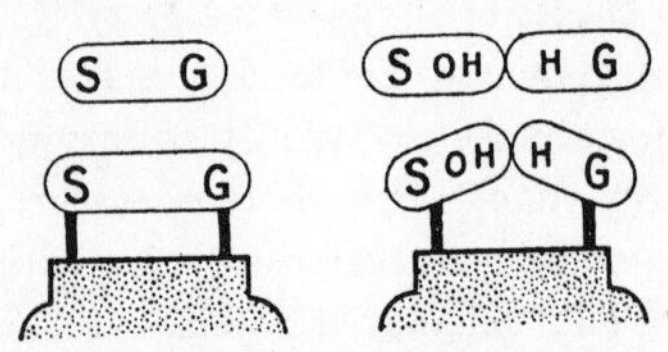

FIG. 35.—Suggested mode of activity of a hydrolytic enzyme.

It is also possible that the enzyme activates the H_2O molecules. The most general theory of enzyme action so far propounded is that of Battelli and Stern [1922], who regarded the activation of H_2O as common to all enzyme actions, hydrolytic reactions and oxidation-reductions included. This is a development of Traube's views. With regard to the former it could hardly be applied to chlorophyllase, which catalyses the reactions

$$RCO_2—C_{20}H_{39} + C_2H_5OH \rightleftharpoons RCO_2—C_2H_5 + C_{20}H_{39}OH,$$

where R is the chlorophyllid residue. It is, moreover, negatived by Nelson and Schubert's [1928] finding (see Chap. III., p. 30), that after saccharase is saturated with sucrose the velocity of hydrolysis varies with the water concentration. Hence we must regard the activated sucrose molecules united with the enzyme as being bombarded by water molecules. Moreover, the evidence is now fairly strong that enzyme-catalysed oxidation-reductions include both direct H transfers and O_2 activation.

The theory has been held that enzyme action is entirely " due to an increase of active mass, owing to concentration on the active surface " [Bayliss, 1925]. Bayliss, it may be added, does not seem to have held this view in its pure form.

Quastel's Theory of Dehydrogenases.

The theory of activation in catalysed oxidations is, of course, extremely controversial. It is not proposed to discuss the various theories which have been, or are, held regarding the mechanism of oxidation. The view here adopted is that certain catalysts, for example, succinoxidase, activate their substrates in such a way that they readily give up a pair of H atoms to a hydrogen acceptor. Conversely they must render the reduced products (e.g. fumaric acid) capable of acting as hydrogen acceptors more readily than usual.

This point of view, due to Wieland, has been developed by Quastel and his colleagues [Quastel, 1926]. He examined the behaviour of 103 substances with *Bacillus coli*, and found that fifty-six were activated. He gives strong reasons for the belief that activation occurs at the cell surface. Associated with certain groups in the cell membrane there exist (on his theory) electric fields of varying strength. The stronger fields should be associated with unstable groupings. They will be superimposed on a general field due largely to selective permeability of the membrane to ions.

If one of these active centres unites with a molecule, the field will tend to divert the electrons and protons of the molecule from their normal orbits. The effect will of course depend on the structure of the molecule concerned. If, however, the molecule, due either to ionization or to the presence of a double bond, is short of protons, they will be able to migrate to an appreciable extent. Thus unsaturated compounds will pass into the following activated forms :—

$$R_1CH{=}CHR_2 \rightarrow R_1\check{C} - CH_2R_2$$

$$R_1CH{=}O \rightarrow R_1\check{C} - OH$$

$$R_1CH{=}N - R_2 \rightarrow R_1\check{C} - NHR_2$$

and the activated forms will be hydrogen acceptors, or in some cases, oxygen acceptors. Similarly an ionized acid will be converted into a hydrogen donator :—

$$
\begin{array}{ccccccccc}
R & & R & & R & & R & & R \\
| & & | & & | & & | & & | \\
CH_2 & \rightleftharpoons & CH_2 & \rightleftharpoons & C\langle{}_{H} & \rightleftharpoons & C\langle + H & \rightleftharpoons & CH \\
| & & | & & | & & | & & \| + H \\
CH_2 & & C\langle{}_{H} & & CH_2 & & CH_2 & & CH \\
| & & | & & | & & | & & | \\
COO/ & & COOH & & COOH & & COOH & & COOH \\
 & & & & \text{(Activated)} & & \text{(Activated)} & &
\end{array}
$$

This theory, combined with the theory as to the effects on the proton of various radicles, enables a successful prediction of the effects of *Bacillus coli* on numerous compounds.

On the other hand, it appears to fail in certain cases. Thus fumaric acid can be activated so as to form succinic, malic, or aspartic acids, by the addition of 2H, H_2O, and NH_3 respectively. The hypothetical HOOC—$\check{C}$—CH_2—COOH should be able to form all three. But succinodehydrogenase, which catalyses the acceptance of 2H, is certainly distinct from fumarase, which catalyses the acceptance of H_2O [Clutterbuck, 1928]. And both succinodehydrogenase and fumarase are distinct from aspartase, which catalyses the union with NH_3 [Woolf, 1929]. The theory will probably be capable of development so as to admit a consideration of the orientation of the substrate molecule, and the finer structure of the activating field.

It may be presumed that the same field not only activates the substrate molecules but attracts them. This is suggested by the following consideration. Table V. (p. 35) shows that, on the whole, the enzymes catalysing oxidation and reduction possess a greater affinity for their substrates than those which catalyse hydrolyses. It is reasonable to suppose that oxidations, which do not proceed in the absence of specific catalysts at temperatures where hydrolyses do so proceed, require a higher energy of activation, and that this is provided at least in part by the relatively large energy of combination of enzyme and substrate.

This view is supported by a consideration of xanthine-oxidase, which activates two types of substrate (see Chaps. III., VI.). Aldehydes are easily oxidized at ordinary temperatures by a variety of reagents, and the enzyme activates them, although its affinity for them is small. (The slow velocity of oxidation is, at least in part, due to the fact that even in 3 per cent. solution, only a small fraction of the enzyme is united with substrate.) Clearly a relatively small amount of energy is needed to activate an aldehyde. On the other hand, hypoxanthine is hardly affected by such a drastic oxidizing agent as boiling HNO_3. But the enzyme has an enormous affinity for it, which may provide the large amount of energy needed for its activation, since a certain parallelism between the free and total energies of enzyme-substrate combination is to be expected. It is shown on theoretical grounds on page 32 that an enzyme's affinity for its substrate must affect the velocity of reaction, but the data are nowhere adequate for a full test of the theory.

It will be clear from the above discussion that we understand activation far less than specificity. This is largely due to the fact that the study of temperature coefficients, which has been so valuable in the chemistry of non-enzymatic reactions, has not been pushed far enough in the case of enzymatic reactions.

Classification of Enzymes.

It is clear that enzymes should be classified by the reactions which they catalyse. In general it is sufficient to indicate the substrate. Sometimes, however, the same substrate may be activated in several ways; thus H_2O_2 is converted into a powerful oxidizing agent by peroxidase, but destroyed by catalase. The most satisfactory terminology is to add the suffix " -ase " (from diastase) to the substrate. On the other hand, there are a large number of different proteases, and it seems better to keep the words pepsin and trypsin for the present, rather than use such a word as pepsinase, which should mean an enzyme attacking pepsin. In the case of the non hydrolytic enzymes, it is usual to state their action, as in the words dehydrogenase and mutase.

Oppenheimer [1925] denotes the enzymes which catalyse hydrolysis and the reverse process as hydrolases. This is not absolutely logical, as chlorophyllase, which catalyses alcoholysis, is clearly analogous to the hydrolases, and Oppenheimer includes it among them. He uses the word desmolases for enzymes breaking the C—C bond, and oxydoreducases for those catalysing oxidation and reduction. While a rational classification of the hydrolases is possible, the remaining " enzymes " are generally mixtures of a number (e.g. zymase) often including crystalloidal catalysts.

Of the enzymes listed in Table XX., an account has been given of the majority elsewhere, especially in Chapter VI. A number of

TABLE XXI.

Enzyme.	*Substrate.*	*Synonyms.*
1. Enzymes Hydrolysing Esters.		
Sulphatase	Phenolic sulphates	
Phosphatase	Phosphoric esters, ? pyrophosphates and their esters	Nucleotidase, ? phytase.
Lipase	Organic esters	Esterase, steapsin.
Tannase	Tannins	? Lipase.
Chlorophyllase	Chlorophyll	
Lecithase	Lecithin	? Lipase + phosphatase.
Pectase	Pectin	? Lipase.
2. Enzymes Hydrolysing Osides and Polysaccharides.		
Arabinosidase [1]	α-*l*-arabinosides	
Nucleosidase	Nucleosides	
Primeverase	Primeverose	Xylosidase.
Maltase	Maltose	
α-glucosidase	Maltose and other α-*d*-glucosides	Gaultherase, maltase.
Trehalase	Trehalose	
Glucosaccharase	Saccharose, etc.	Invertase, sucrase.
β-glucosidase	β-glucosides	Prunase, emulsin, etc.
Gentiobiase	Gentiobiose	May be specific enzymes, or aspects of β-glucosidase.
Cellobiase	Cellobiose	
Amygdalase	Amygdalin (partly)	
Sinigrinase	Sinigrin	Myrosin.
Fructosaccharase	Saccharose, raffinose	Invertase, sucrase, raffinase.
α-mannosidase	α-mannosides	
Melibiase	Melibiose	α-galactosidase, ? galactoraffinase.
Lactase	Lactose, β-galactosides	β-galactosidase.
Rhamninorhamnase	Rhamninose	
Robinase	Robinose	
Indimulsin	Indican	
Isatase	Isatin glucoside	
Rhamnodiastase	Xanthorhamnin, etc.	Rhamninase.
Erythrozym	Ruberythric acid	
Amylase	Starch and glycogen	Diastase, amylopsin.
Cellulase	Cellulose and lichenin	Cytase, lichenase.
Polyamylase	Polyamyloses	
Mannanase	Mannans	Seminase.
Inulase	Inulin	
Chitinase	Chitin	
Pectinase	Pectin	
3. Enzymes Hydrolysing the C—N Linkage.		
Urease	Urea	
Histozym	Hippuric acid, etc.	
Arginase	Arginine, etc.	
Asparaginase	Asparagine (? other amides)	? Erepsin.
Allantoinase	Allantoin	
Guanase	Guanine	Guanine deaminase.
Adenase	Adenine	Adenine deaminase.
Peptidases	Peptides	Erepsin, (trypsin).
Trypsin	Proteins, peptides, etc.	Trypsin-kinase, tryptase.
Kathepsin	Proteins	Autolytic protease.
Papain	Proteins, etc.	
Bromelin	,, ,,	
Pepsin	Proteins	Peptase.
Rennin	Phosphoproteins	Chymosin, chymase.
? Thrombin	Fibrinogen	Thrombase, serozym.

[1] (Bridel et Béguin, 1926.)

TABLE XXI.—*continued.*

Enzyme.	*Substrate.*	*Synonyms.*
4. Enzymes Inserting H_2O without Hydrolysis.		
Fumarase	Fumaric acid	
Glyoxalase	Substituted glyoxals	Ketonaldehydemutase.
5. Enzyme Removing NH_3 without Hydrolysis.		
Aspartase	*l*-Aspartic acid.	
6. Enzymes Attacking the C—C Bond.		
Oxynitrilase	Mandelonitrile	Benzcyanase, emulsin
Carboxylase	α-keto-acids	
Carboligase	Acyloin [1]	
Zymase	Hexoses, etc.	(A mixture of enzymes.)
Myozymase, etc.	,, ,,	,, ,, ,,
7. Enzymes Activating H_2O_2.		
Catalase	H_2O_2	
Peroxidase	H_2O_2	
8. Enzymes Activating other Oxidants.		
Oxygenase	O_2	Atmungsferment, ? Polyphenol oxidase, indophenol oxidase.
Nitrate activator	Nitrates	
Chlorate ,,	Chlorates	
9. Dehydrogenases (also Described as Oxidase, Oxydone, Reductase, Perhydridase, Oxido-Reductase, Redoxase, Hydrogen-Transportase, etc.).		
Formic dehydrogenase	Formic acid	
Succinic ,,	Succinic ,,	Succinoxydone, Succinoxidase.
Sugar ,,	Hexoses	
Aldehyde ,,	Aldehydes	
Lactic ,,	α-hydroxy-acids	
Citric ,,	Citric acid	
Xanthine ,,	Xanthine, etc., aldehydes	Schardinger enzyme.
Alcohol ,,	Ethanol	Alcoholase
Uricase, etc., etc.	Uric acid	Uric oxidase.
10. Unclassified Oxidases (Some Probably Mixtures).		
Tyrosinase	Tyrosine, etc., O_2	
Dopa oxidase	3, 4-di-hydroxy-phenylalanine, O_2	
Tyramine oxidase	Tyramine, O_2	
Indophenol oxidase	*p*-phenylenediamine, O_2	? Oxygenase.
Laccase	Guaiacol, etc., O_2	? Includes peroxidase.
Mutase	Aldehydes	Aldehydase, oxidoreductase.

doubtful enzymes have not been included. While tissue phosphatases attack a wide variety of substrates, e.g. nucleotides, it is not certain whether they hydrolyse phytin or whether a special phytase exists. Nucleosidase, which hydrolyses purine β-*d*-ribosides, does not attack αβ-methyl-*d*-riboside or α- or β-methyl xyloside [Levene, Jacobs, and Medigreceanu, 1912]. Hence, even if the nucleosides

[1] Only synthesis of this substance from aldehyde so far described.

are xylosides [Robinson, 1927], nucleosidase must be analogous to the doubly specific maltase, cellobiase, etc., rather than to the relatively unspecific β-glucosidase of emulsin. Primeverose is glucose xyloside. Hence, primeverosidase, which hydrolyses primeverosides [Bridel, 1926, 1] is probably β-glucosidase, while primeverase, which hydrolyses primeverose, is probably a xylosidase [Bridel, 1926, 2]. While a certain amount of order has been reached (see Chap. VI.) in the study of the enzymatic hydrolysis of glucosides and polysaccharides derived from the hexoses, that of the methyl-pentose derivatives is still very obscure. No enzyme has yet been found which will hydrolyse rhamnosides such as quercitrin. On the other hand, the sugars rhamninose [Bierry, 1909] and robinose [Charaux, 1926], both of which consist of two rhamnose and one galactose residues, can be hydrolysed by enzymes derived from the snail and acacia respectively. *Rhamnus utilis* yields an enzyme or enzyme mixture rhamnodiastase, which hydrolyses osides derived from primeverose, rhamninose, rutinose [Bridel et Charaux, 1926] and robinose [Charaux, 1926], but does not act on the sugars. Until the structure of these sugars is determined the interpretation of these results must remain uncertain. Some of the more complicated glucosides, such as those of indoxyl [Hazewinkel, 1900], isatin [Beyerinck, 1900], and alizarin [Schunck, 1854], which are wholly or largely resistant to emulsin, are hydrolysed by special enzymes.

The specificity of deaminating enzymes is still in some confusion. However, Jones [1914] brings forward evidence for the separate existence of deaminases acting on guanine, adenine, guanosine, and adenosine. Allantoinase [Fosse and Brunel, 1929] is almost unique in opening a ring :—

$$NH_2\text{—CO—NH—}\underset{\displaystyle\text{CH—NH}}{\overset{\displaystyle\text{CO—NH}}{|}}\!\!\!>\!\text{CO} + H_2O = NH_2\text{—CO—NH—}\overset{\displaystyle\text{COOH}}{\underset{\displaystyle\text{CH}}{|}}\text{—NH—CO—NH}_2.$$

The proteases may be classified according to their optimum pH. They probably fall into three groups. Pepsin acts at the pH at which the proteins have a maximum positive charge. A group of intracellular proteases, including papain, bromelin, yeast protease, rennin and kathepsin, the autolytic protease of mammalian tissues [Waldschmidt-Leitz and Deutsch, 1927] act best at the isoelectric point. This fact is commonly obscured by the compresence of erepsin. Hence, if the rate of amino nitrogen production is measured the optimum lies on the alkaline side of the isoelectric point, or there may be two optima. Finally, trypsin has an optimum well on the alkaline side of the isoelectric point. It will ultimately be possible to classify the proteases according to their range of substrates, but this is as yet only possible in a few cases (see Chap. VI.). Probably avian keratinase [Stankovic et al., 1929] which has an alkaline optimum, and liberates amino-acids from keratin, will be found to fall into a special class. The physiological evidence suggests that it leaves behind it a part of the keratin molecule, which is available for feather growth.

The desmolases, or enzymes attacking the C—C bond are still little understood. Zymase, for example, is clearly a mixture (see Harden's *Alcoholic Fermentation* in this series). Carboxylase appears to be a specific component, but we do not know whether the following processes are performed by different enzymes:—

(*a*) Phosphorylation of glucose.

(*b*) Breaking down of the hexose molecule probably with simultaneous formation of hexose-diphosphate.

(*c*) Mutase action, e.g.,

$$2CH_3{-}COH = CH_3{-}CO{-}O{-}C_2H_5 \text{ (or } CH_3{-}COOH + HO{-}C_2H_5)$$

and probably

$$CH_3{-}CO{-}COH + CH_3{-}COH + H_2O = CH_3{-}CO{-}COOH + C_2H_5OH.$$

(*d*) Acyloin synthesis from aldehyde, e.g.,

$$2CH_3{-}COH = CH_3{-}CO{-}CHOH{-}CH_3,$$

although (*b*) and (*c*) appear to require the same cozymase (see Chap. VII.).

Some of the catalysts of glycolysis, for example, that of the red blood corpuscles, have resisted all attempts to isolate them from the cell. Those of yeast and muscle have been prepared cell-free. Relatively little effort has been made to isolate others, e.g. those of lactic acid bacteria, brain, etc. Until the various components of the zymase system have been studied apart, it is idle to attempt their classification.

Oxidases.

A catalyst of oxidation and reduction is generally regarded either as activating a certain type of molecule (which is either oxidized or reduced in the process catalysed), or in some cases (e.g. glutathione) as forming an intermediate compound. While in complex systems there is clear evidence of activation of both oxidizing and reducing molecules simultaneously (e.g. in the system xanthine-oxidase + peroxidase (Thurlow, 1925), it is not yet shown that this can be done by a single enzyme, though the possibility remains open. It is, therefore, often convenient to classify oxidases according as they activate the oxidizing molecule (O_2, H_2O_2, nitrate, etc.) or the reducing molecule (most organic compounds). But this classification does not take account of the reversibility of enzyme reactions. Often this does not matter. Thus a calculation from thermodynamic data shows that catalase would only act synthetically to an appreciable extent under an O_2 pressure of many billions of atmospheres. It would, therefore, be perverse, if logical, to describe it as water oxidase. Similarly, if the reactions catalysed by peroxidase can be represented by: $H_2O_2 + 2H \rightleftharpoons 2H_2O$, peroxidase might be regarded as water dehydrogenase, although no case of such reversed action has been observed. But the enzyme which catalyses the oxidation or rather dehydrogenation of succinic to fumaric acid can also catalyse the reverse process not only in vitro [Quastel and Whetham, 1924] but also as a source of energy for the living organism [Quastel, Stephenson, and Whetham, 1925].

The question of specificity waits on the isolation of the enzymes. There is no doubt at all that peroxidase and catalase are definite chemical individuals. On the other hand, very little is known regarding the activators of O_2 on the one hand, nitrates and chlorates on the other. Willstätter [1926] does not regard Warburg's Atmungsferment as an enzyme at all. Only a few of the activators of organic molecules have been isolated from other oxidases. It is most convenient to regard these as dehydrogenases, though they may act reversibly. Their specificity is already known to vary with their source. Thus milk aldehyde-oxidase activates xanthine, potato aldehyde oxidase does not. It may also vary with their environment. Thus Quastel and Wooldridge [1927, 1928] find that while a soluble enzyme obtained by the autolysis of *Bacillus coli* activates lactic acid, causing the reduction of methylene blue as rapidly as the original organism, it does not activate succinic acid, but there is strong ground for believing that this substrate is activated by the lactic acid enzyme when this forms part of the cell (see Chap. VI.).

Mixed Oxidase Systems.

The system of catalysts responsible for oxidation and reduction in a living cell may include :—

1. Activators of O_2 (more rarely —NO_3', ClO_3', etc.). These include complex organic molecules, e.g. hæmin and its compounds [Krebs, 1928], and salts of iron or other heavy metals [Bertrand, 1897; Euler and Bolin, 1909].

2. Peroxidases, activating H_2O_2. These may be enzymatic or heat-stable.

3. Dehydrogenases producing H_2O_2 from O_2, e.g. xanthine-oxidase and tyramine-oxidase.

4. Dehydrogenases requiring an O_2 activator if they are to reduce O_2 directly, e.g. succinoxidase, bacterial lactic dehydrogenase [Stephenson, 1928].

5. Catalase, which probably acts mainly, if not wholly, as a safety-valve, though the possibility of linked reactions utilizing the energy of the reaction $2H_2O_2 = 2H_2O + O_2$ cannot be ignored.

6. Autoxidizable substances readily reduced in the cell, e.g. glutathione, hæmatins, and cytochrome *b* [Keilin, 1929].

7. Intermediate compounds oxidized by an enzyme, reduced by a reactant, e.g. the catechol derivative described by Onslow [1920] in many plants, and St. György's [1928] hexuronic acid which is oxidized by peroxidase in presence of catechol derivatives, and reduced by glutathione and other hydrogen donators. Keilin's [1929] cytochrome *a* and *c* probably fall into this class. Their oxidation is inhibited by CO, however CO does not combine with them, but probably acts by inhibiting a catalyst of class 1.

8. Possibly enzymes which are specific both for a reducing and an oxidizing substrate. Tyrosinase, for example, acts on tyrosine, or on certain mixtures containing phenolic and amino groups, e.g. *p*-cresol + alanine or glycine, causing their oxidation by O_2 [see summary of literature by Raper, 1928], but does not catalyse methylene blue reduction by tyrosine, though it does catalyse the reduction of methylene blue by glycine + *m*-cresol or *p*-cresol [McCance, 1925]. It, therefore, seems plausible that tyrosinase activates O_2. The O_2 activator may or may not be separable from the tyrosine activator.

Perhaps some of the dehydrogenases of class 4 really belong here, e.g. succinoxidase may possess an O_2 activating group which is inhibited by HCN. But no such explanation can apply to Stephenson's lactic dehydrogenase.

In the above classification the catalysts of group (3) which produce H_2O_2 from O_2 correspond to Stern's [1927] oxydases; those of groups (4) and (8), which do not produce H_2O_2, to her oxydones. The correspondence is perhaps not quite exact, and any attempt at classification is clearly provisional.

A large number of oxidases are not enzymes in the strict sense of the word, and some can act in several different ways. Thus cytochrome and hæmatin appear to be the principal "peroxidases" in animal [Keilin, 1929], though not plant cells, as well as intermediates in oxidations not involving H_2O_2. Various "polyphenol-oxidases" are heat-stable. Thus Euler and Bolin [1908-1909] and Wieland and Fischer [1926] have obtained heat-stable catalysts of this class from plants. Some of these consist of salts, and their action can be imitated by such mixtures as calcium glycollate + manganous acetate. There are also thermolabile and presumably enzymatic polyphenol-oxidases, such as the quinol and catechol oxidases of Wieland and Sutter [1928] and the yeast indophenol and potato catechol oxidases of Keilin [1929]. In many cases it is not clear whether they can act anaerobically or not. However, yeast indophenol oxidase and potato catechol oxidase are inhibited by CO [Keilin, 1929], so they are clearly activators of O_2, or at least systems in which that activation is the limiting factor. Since CO and HCN have much the same effect on yeast indophenol oxidase as on O_2 uptake by the living cell, it appears that most of the O_2 respired is activated by indophenol oxidase, which is therefore to be identified with Warburg's Atmungsferment. Unfortunately very little is known as to its properties. Although Bach originally used the word "oxygenase" in a different connexion, this appears to be the most correct appellation for the Atmungsferment.

The dehydrogenases are very hard to classify, as most dehydrogenations so far studied are performed by bacteria or tissues [summary by Bernheim, 1928, 2]. Some of them have been isolated, but as pointed out on pages 61, 62 this may alter their properties. The list given is certainly quite incomplete. Moreover, the range of specificities varies as we saw in an unpredictable manner.

Note on Nomenclature.

Duclaux [1883] introduced the custom of designating an enzyme by the substrate on which its action was first observed, and adding the suffix "-ase." This is on the whole convenient, but is not fully adequate for the following reasons :—

1. It is desirable, if possible, to use a name covering the group of substances hydrolysed, e.g. β-glucosidase, rather than salicinase or gentiobiase. The term gentiobiase could then be reserved for an enzyme hydrolysing gentiobiose alone, if such exists.

2. Within a group, say of proteases, it is necessary to distinguish different types, e.g. pepsin and trypsin. There can be little advantage in calling trypsin tryptase in violation of Duclaux's rule.

3. The same substrate may be transformed in different ways by different enzymes, e.g. fumaric acid may be induced to unite with H_2O, NH_3, or 2H. Hence different names must be used.

4. There is often great uncertainty as to the individuality of a given enzyme. We do not know, for example, whether some aldehyde oxidases can also function as aldehyde-mutases, nor how many different enzymes are contained in emulsin.

For this reason no hard and fast rules are desirable. Duclaux's terminology should be followed as far as possible, but the time has not yet come when any definite system of terminology has much chance of final acceptance. This is especially the case with the catalysts of oxidation and reduction. Before the reversible nature of enzyme action was clearly recognised the suffix "-ese" was suggested for synthetic enzymes. This has, for obvious reasons, found little application, though it is convenient to call the enzyme synthesizing (2-5) fructose diphosphate as part of a more complex reaction "phosphatese."

BIBLIOGRAPHY.

Papers marked with an asterisk are collected in Willstätter's "Untersuchungen über Enzyme" (Springer, 1928).

Page of Text on which reference is made.

E. ABDERHALDEN and H. BROCKMANN (1928, 1), *Studien über den fermentativen Abbau von Polypeptiden verschiedener Zusammensetzung* . . 113
Fermentforschung, **9**, 446.

E. ABDERHALDEN and H. BROCKMANN (1928, 2), *Zur Frage der die Hydrolyse von Polypeptiden einleitenden spezifischen Bindungsart von Substrat und Fermentkomplex* 113
Fermentforschung, **10**, 159.

E. ABDERHALDEN, G. CAEMMERER, and L. PINCUSSEN (1909), *Zur Kentniss des Verlaufs der fermentativen Polypeptidspaltung. VII. Mitteilung* . . 66
Zts. physiol. Chem., **59**, 293.

E. ABDERHALDEN and R. FLEISCHMANN (1928), *Studien über den fermentativen Abbau von Polypeptiden verschiedener Zusammensetzung und deren Verhalten gegenüber n-Alkali* 112, 113
Fermentforschung, **9**, 524.

E. ABDERHALDEN and A. FODOR (1916), *Forschungen über Fermentwirkung. I. Mitteilung. Studien über den fermentativen Abbau von Polypeptiden* 23
Fermentforschung, **1**, 533.

E. ABDERHALDEN and H. HANDOWSKY (1921), *Beiträge zur Frage des Einflusses der Struktur und Konfiguration des Substrates (Polypeptide) auf die Fermentwirkung* 111
Fermentforschung, **4**, 316.

E. ABDERHALDEN and J. HARTMANN (1927), *Über das Verhalten von dl-Alanyl-δ-aminovaleriansaüre gegenüber Polypeptidasen* 112
Fermentforschung, **9**, 199.

E. ABDERHALDEN and W. KÖPPEL (1928, 1), *Studien über den fermentativen Abbau von Polypeptiden, an deren Aufbau l-Oxyprolin beteiligt ist* 113
Fermentforschung, **9**, 439.

E. ABDERHALDEN and W. KÖPPEL (1928, 2), *Vergleichende Studien über den fermentativen Abbau von Polypeptiden durch Erepsin und Trypsin-Kinase* 113
Fermentforschung, **9**, 516.

E. ABDERHALDEN, H. PIEPER, and R. TATEYAMA (1926), *Das Verhalten von dl-Leuzyl-γ-aminobuttersaure gegenüber Hefemazerationssaft* . . . 112
Fermentforschung, **8**, 579.

E. ABDERHALDEN and F. REICH (1928), *Studien über den Einfluss von n-Alkali, von Erepsin und von Trypsin-Kinase auf Polypeptidartige Verbindungen, an deren Aufbau β-Alanin beteiligt ist* 112, 113
Fermentforschung, **10**, 173.

E. ABDERHALDEN and E. ROSSNER (1928), *Versuche über die Einwirkung von Erepsin und Trypsin auf d-Glutaminsaure enthaltende Polypeptide* . 113
Fermentforschung, **9**, 501.

E. ABDERHALDEN and N. SCHAPIRO (1927), *Über das Verhalten von l-Leucyl-glycyl-l-tyrosin und d-Leucyl-glycyl-l-tyrosin gegenüber Hefemazerationssaft, Pankreas- und Darmsaft* 111
Fermentforschung, **9**, 234.

Page of Text on which reference is made.

E. ABDERHALDEN and E. SCHWAB (1928, 1), *Weiterer Beitrag zur Frage der spezifischen Einstellung der Polypeptidasen* 111, 113
Fermentforschung, **9**, 501.

E. ABDERHALDEN and E. SCHWAB (1928, 2), *Weiterer Beitrag zur Frage der spezifischen Wirkung von Erepsin und Trypsin-Kinase* . . . 113
Fermentforschung, **10**, 179.

E. ABDERHALDEN and E. SCHWAB (1930), *Studien über das Verhalten von Polypeptiden an deren Aufbau Thyroxin beteiligt ist, gegenüber Fermenten, und Prüfung ihres biologischen Verhaltens* . . . 113, 115
Fermentforschung, **11**, 164.

E. ABDERHALDEN and H. SICKEL (1928), *Weitere Studien über den fermentativen Abbau von Polypeptiden verschiedener Zusammensetzung* . . 113
Fermentforschung, **9**, 463.

E. ABDERHALDEN and A. WEIL (1920), *Studien über Lipasewirkung* . . 102, 103
Fermentforschung, **4**, 76.

E. ABDERHALDEN and W. ZEISSET (1929), *Weiterer Beitrag zur Frage den Einheitlichkeit des Trypsinkomplexes. III. Mitteilung* . . . 114
Fermentforschung, **10**, 481.

E. ABEL (1922), *Über die catalatische Wirkung der Peroxydase* . . . 18
Zeit. Elektrochem., **28**, 489.

G. S. ADAIR (1928), *A theory of partial osmotic pressure and membrane equilibria, with special reference to the application of Dalton's law to hæmoglobin solutions in the presence of salts* 29
Proc. Roy. Soc., A., **120**, 573.

L. ADLER (1915), *Uber die Phosphatasen im Malz* 15
Bioch. Zts., **70**, 1.

AHLGREN (1925), *Zur Kenntnis der Tierischen Gewebeoxydation* . . . 37
Skand. Arch. Physiol., **47** (Supplement), 44.

N. ALWALL (1928), *Eine Methode um fumarasefreie Succinodehydrogenase zu bekommen* 126
Skand. Arch. Physiol., **54**, 1.

L. AMBARD (1923), *De l'amylase. Liaison du ferment et des substances qu'il digére* 63
Bull. Soc. Chim. Biol., **5**, 693.

G. V. ANREP and R. K. CANNAN (1923), *The concentration of lactic acid in the blood in experimental alkalæmia and acidæmia* 3
J. Physiol., **58**, 244.

E. F. ARMSTRONG (1904), *Studies on enzyme action. III. The influence of the products of change on the rate of change by sucroclastic enzymes* . . 9, 36
Proc. Roy. Soc., B., **73**, 516.

E. F. ARMSTRONG (1905), *Studies on enzyme action. VII. The synthetic action of acids contrasted with that of enzymes. Synthesis of maltose and isomaltose* 4
Proc. Roy. Soc., B., **76**, 592.

E. F. ARMSTRONG (1924), "*The Simple Carbohydrates and the Glucosides.*" (This series) 4

H. E. ARMSTRONG and E. ORMEROD (1906), *Studies on enzyme action—Lipase, II.* 103
Proc. Roy. Soc., B., **78**, 376.

H. E. ARMSTRONG, E. F. ARMSTRONG, and E. HORTON (1908), *Studies on enzyme action. XII. The enzymes of emulsin* 93
Proc. Roy. Soc. B., **80**, 321.

S. ARRHENIUS (1907), *Immunchemie* 69, 91

K. ASAKAWA (1929), *Uber die fermentative Spaltung der verschiedenen Phosphorsaüreester* 15
J. Biochem., **11**, 143.

S. J. AULD (1908), *The hydrolysis of amygdalin by emulsin. Part I.* . . 36
Journ. Chem. Soc., **93**, 1251.

Page of Text on which reference is made.

O. T. Avery and G. E. Cullen (1920), *Studies on the enzymes of Pneumococcus.*
II. Lipolytic enzymes. (*Esterase*) 15
J. Exp. Med., **32**, 571.
III. Carbohydrate splitting enzymes. (*Invertase, amylase, inulase*) . 16
J. Exp. Med., **32**, 583.

D. Bach (1928). *Les conditions d'action de l'asparaginase* 17, 36
C. R. Ac. Sc., **187**, 955.

D. Bach (1929), *Sur quelques conditions d'action de l'urease de l'Aspergillus niger* 17
C. R. Soc. Biol., **100**, 831.

A. Bach and R. Chodat (1904), *Untersuchungen über die Rolle der Peroxyde in der Chemie der lebenden Zelle. VIII. Über die Wirkungsweise der Peroxydase* 77
Ber., **37**, 1342.

A. Bach, W. Engelhardt, and A. Samyslow (1925), *Über die Rolle der Begleitstoffe bei der Immunisierung mit Invertasepraparaten* . . 165
Bioch. Zts., **160**, 261.

E. Bamann (1929), *Über die Konfigurationsspezifität der Leber-Esterase verschiedener Tiere, und ihre Abhängigkeit von der Substrat-Konzentration* 33, 35, 88, 104
Ber., **62**, 1538.

E. Bamann and M. Schmeller (1929), *Zur Kinetik der Esterhydrolyse durch Enzyme* 35, 85
Zts. physiol. Chem., **183**, 149.

H. P. Barendrecht (1920), *L'uréase et la théorie de l'action des enzymes par rayonnement* 2, 28
Rec. Trav. Chim., **39**, 2.

H. P. Barendrecht (1924), *Saccharase und die zweite Wirkungsart der Wasserstoffionen* 28
Bioch. Zts., **151**, 363.

J. la Barre (1925), *On the inactivation of atropine by rabbit serum* . . 102
Journ. Pharm. and Exp. Ther., **26**, 259.

F. Battelli and L. Stern (1911), *Zur Kentniss des Pneins* 145
Bioch. Zts., **33**, 315.

F. Battelli and L. Stern (1912), *Die Oxydationsfermente* 9
Ergebn. d. Physiol., **12**, 96.

F. Battelli and L. Stern (1922). *Oxydations et réductions fermentatives* 168, 182
C. R. Ac. Sc. Geneve., **37**, 65.

W. M. Bayliss (1904), *The kinetics of tryptic action* 36, 66
Arch. Sci. Biol., **11**, 261.

W. M. Bayliss (1913), *Researches on the nature of enzyme action. III. The synthetic action of enzymes* 2, 128
J. Physiol., **46**, 236.

W. M. Bayliss (1925), "*The Nature of Enzyme Action.*" (This series) . 3, 183

W. M. Bayliss and E. Starling (1904), *The proteolytic activities of the pancreatic juice* 64
J. Physiol., **30**, 601.

H. Bechhold and L. Keiner (1927), *Trennung von Trypsin und Enterokinase durch Ultrafiltration* 142, 179
Bioch. Zts., **189**, 1.

F. Bernheim (1928, 1), *The aldehyde oxidase of the potato*. 18, 126
Bioch. J., **22**, 344.

F. Bernheim (1928, 2), *The separation of citric acid dehydrase from liver and of the lactic acid dehydrase from yeast* 34, 37, 59, 126, 192
Bioch. J., **22**, 1178.

F. Bernheim and M. Dixon (1928), *Studies on xanthine oxidase. X. The action of light* 72
Bioch. J., **22**, 113.

Page of Text on which reference is made.

N. J. Berrill (1929), *Digestion in ascidians and the influence of temperature* 16
Brit. Journ. Exp. Biol., **6**, 275.

G. Bertrand (1897), *Sur l'intervention du manganèse dans les oxidations provoquées par la laccase* 190
C. R. Ac. Sc., **124**, 1355.

G. Bertrand and A. Compton (1911), *Influence de la réaction du milieu sur l'activité de la cellase. Nouveau caractère distinctif d'avec l'émulsine* . 16
C. R. Ac. Sc., **153**, 360.

M. W. Beyerinck (1900), *On Indigo-fermentation* 188
Kon. Akad. Wet. Amst., Proceedings of the section of sciences, **2**, 495.

Berzelius (1837), *Lehrbuch der Chemie.* Pages 6-20 8

H. Bierry (1909), *Dédoublement diastasique du rhamninose* . . . 188
C. R. Soc. Biol., **66**, 738.

H. Bierry (1909), *Invertines et lactases animales, leur spécificité* . . 95
C. R. Ac. Sc., **148**, 949.

H. Bierry and J. Giaja (1908), *Sur le dédoublement diastasique du lactose, du maltose, et de leurs dérivés* 95
C. R. Ac. Sc., **147**, 268.

A. V. Blagoveschenski and A. N. Bieloserski (1925), *The specific action of plant ferments. II. The specific conditions of action of leaf peptases* 18
Bioch. J., **19**, 355.

A. V. Blagoveschenski and N. I. Sossiedow (1925), *The specific action of plant ferments. I. The specific conditions of action of leaf invertase* 16
Bioch. J., **19**, 350.

A. V. Blagoveschenski and N. I. Sossiedow (1927), *The specific action of plant enzymes. III. The specific conditions of action of leaf salicinases* 16
Bioch. J., **21**, 1206.

W. R. Bloor (1912), *Carbohydrate esters of the higher fatty acids. II. Mannite esters of stearic acid* 102
J. Biol. Chem., **11**, 141.

M. Bodansky (1919), *A note on the determination of catalase in blood* . 18
J. Biol. Chem., **40**, 127.

E. Bourquelot (1917), *La synthèse biochimique des glucosides d'alcools. IV. Galactosides d'alcools* 96
Ann. Chim., **7**, 153.

E. Bourquelot and A. Aubry (1917), *Synthèse biochimique d'un deuxième galactobiose à l'aide de l'émulsine* 95
J. Pharm. et Chim., **15**, 273.

E. Bourquelot (1914), *Synthèse biochimique des glucosides et des polysaccharides. Réversibilité des actions fermentaires* 2, 82, 97
J. Pharm. et Chim., **10**, 361, 393.

E. Bourquelot (1898), *Sur la physiologie du gentianose, son dédoublement par les ferments solubles* 100
C. R. Ac. Sc., **126**, 1045.

E. Bourquelot and M. Bridel (1911), *Action de l'invertine sur les polysaccharides dérivés du lévulose. Application à l'étude du poids moléculaire du verbascose* 99
J. Pharm. et Chim., **3**, 569.

H. Borsook and H. Wasteneys (1925), *The enzymic synthesis of protein. IV. The effect of concentration on peptic synthesis* 2
J. Biol. Chem., **63**, 563.

G. Bredig and P. Müller von Berneck (1899), *Über anorganische Fermente, I.* 6
Zts. physikal. Chem., **31**, 258.

M. Bridel (1926, 1), *Sur la présence, dans l'émulsine des amandes, de deux nouveaux ferments, la primevérosidase et la primevérase* 188
Bull. Soc. Chim. Biol., **8**, 67.

Page of Text on which reference is made.

M. Bridel (1926, 2), *Réflexions sur les diastases et leur spécificité* . . . 188
Bull. Soc. Chim. Biol., **8,** 170.

M. Bridel and Th. Aagaard (1927), *Recherches sur l'hydrolyse diastasique du melezitose at du turanose* 95, 99, 100
Bull. Soc. Chim. Biol., **9,** 884.

M. Bridel and C. Beguin (1926), *Synthése biochimique, a l'aide de l'émulsine des amandes, de l'éthyl-l-arabinoside α* 186
Bull. Soc. Chim. Biol., **8,** 469.

M. Bridel and C. Charaux (1926), *Le produit fermentaire des graines de divers* Rhamnus *ou Rhamnodiastase* 188
Bull. Soc. Chim. Biol., **8,** 35.

G. E. Briggs and J. B. S. Haldane (1925), *A note on the kinetics of enzyme action* 40
Bioch. J., **19,** 338.

A. J. Brown (1902), *Enzyme action* 9, 38
Trans. Chem. Soc., **81,** 373.

H. T. Brown and G. Morris (1895), *Note on the action, in the cold, of diastase on starch-paste* 117
Journ. Chem. Soc., **67,** 309.

E. Buchner (1897), *Alkoholische Garung ohne Hefezellen* 1, 8
Ber., **30,** 117.

E. Buchner and F. Klatte (1908), *Über das Ko-Enzym des Hefepressaftes* 137
Bioch. Zts., **8,** 520.

F. Cajori (1930). In the press. 36, 101

E. M. Case and D. R. McCullagh (1928), *Pancreatic extract in relation to lactic acid formation in muscle* 163
Bioch. J., **22,** 1060.

C. Charaux (1926), *Sur le dédoublement biochemique du Robinoside. Robinose, nouveau triose, provenant de ce dédoublement* 188
Bull. Soc. Chim. Biol., **8,** 915.

A. Clementi (1916), *L'Arginasi come fermento ureogenetico, e la specificita della sua azione deguanidizzante* 107
Arh. di. Fisiol., **14,** 207.

P. W. Clutterbuck (1928), *Succinoxidase. II. Influence of phosphate and other factors on the action of the succinodehydrogenase and the fumarase of liver and muscle* 184
Bioch. J., **22,**, 1193.

O. Cohnheim (1901), *Die Umwandlung des Eiweiss durch die Darmwand* . 107
Zts. physiol. Chem., **33,** 451.

S. W. Cole (1904), *The influence of electrolytes on the influence of amylolytic ferments* 138
J. Physiol., **30,** 202.

H. Colin and A. Cugnac (1926), *Les lévulosanes des Graminées : graminine et triticine* 100
Bull. Soc. Chim. Biol., **8,** 621.

H. I. Coombs (1927), *Studies on xanthine oxidase. IX. The specificity of the system, II.* 123
Bioch. J., **21,** 1259.

A. Croft Hill (1898), *Reversible zymohydrolysis* 9
J. Chem. Soc., **73,** 634.

H. D. Dakin (1903), *The hydrolysis of optically inactive esters by means of enzymes. Part I.* 103
J. Physiol., **30,** 253.

H. D. Dakin (1922), *The action of muscle tissue on fumaric, maleic, glutaconic, and malic acids* 128
J. Biol. Chem., **52,** 183.

Page of Text on which reference is made.

H. D. DAKIN and H. W. DUDLEY (1913, 1), *The action of enzymes on racemised proteins and their fate in the animal body* 111
J. Biol. Chem., **15**, 271.

H. D. DAKIN and H. W. DUDLEY (1913, 2), *On glyoxalase* . . . 3, 127
J. Biol. Chem., **14**, 423.

H. D. DAKIN and H. W. DUDLEY (1913, 3), *Glyoxalase. III. The distribution of the enzyme and its relation to the pancreas* . . . 163
J. Biol. Chem., **15**, 463.

H. D. DAKIN and H. W. DUDLEY (1914), *Glyoxalase, IV.* . . . 127
J. Biol. Chem., **16**, 505.

A. DANILEWSKY (1862), *Über specifisch wirkende Körper des natürlichen und künstlichen pankreatischen Safts* 8
Arch. Path. Anat. und Physiol., **25**, 279.

E. DAWSON, B. PLATT, and J. COHEN (1926), *The hydrolysis of asymmetric esters by lipase* 103
Bioch. J., **20**, 533.

C. DELEZENNE and S. LEDEBT (1922), *Sur la transmission en série du pouvoir protéolytique conféré au suc pancréatique par l'enterokinase* . . 141
C. R. Ac. Sc., **175**, 779.

W. DIETZ (1907), *Über eine umkehrbare Fermentreaktion im heterogenen System. Esterbildung und Esterverseifung* 3
Zts. physikal. Chem., **52**, 279.

M. DIXON (1926), *Studies on xanthine oxidase. VII. The specificity of the system* 123
Bioch. J., **20**, 703.

M. DIXON (1929), *Oxidation mechanisms in animal tissues* . . . 127
Biol. Rev., **4**, 352.

M. DIXON and K. A. ELLIOTT (1929), *The effect of cyanide on the respiration of animal tissues* 156, 157
Bioch. J., **23**, 812.

M. DIXON and K. KODAMA (1926), *On the further purification of the xanthine oxidase* 58, 166, 171-174
Bioch. J., **20**, 1104.

M. DIXON and S. THURLOW (1924), *Studies on xanthine oxidase. II. The dynamics of the oxidase system* . . . 13, 18, 33, 34, 37, 55-59, 66
Bioch. J., **18**, 976.

M. DIXON and S. THURLOW (1925), *Studies on xanthine oxidase. VI. A cell oxidation system independent of iron* 156
Bioch. J., **19**, 672.

C. G. DOUGLAS, J. S. HALDANE, and J. B. S. HALDANE (1912), *The laws of combination of hæmoglobin with CO and O_2* 44, 52
J. Physiol., **44**, 275.

G. DREYER and O. HANSSEN (1907), *Recherches sur les lois de l'action de la lumière sur les glucosides, les enzymes, les toxines, les anticorps* . . 72
C. R. Ac. Sc., **145**, 564.

E. DUCLAUX (1883), *Microbiologie*, p. 141. 193

E. DUCLAUX (1898), *Lois générales de l'action des diastases* . . . 9
Ann. Inst. Past., **12**, 96.

G. S. EADIE (1926), *The effect of substrate concentration on the hydrolysis of starch by the amylase of germinated barley* 34, 36, 42, 43, 68
Bioch. J., **20**, 1016.

G. S. EADIE (1927), *On liver amylase* 16, 36, 139
Bioch. J., **21**, 314.

S. EDLBACHER (1917), *Versuche über Wirkung und Vorkommen der Arginase* 107
Zts. physiol. Chem., **100**, 111.

S. EDLBACHER (1926), *Zur Kentniss des intermediären Stoffwechsel des Histidins* 17
Zts. physiol. Chem., **157**, 106.

Page of Text on which reference is made.

S. EDLBACHER and P. BONEM (1925), *Beiträge zur Kenntnis des Arginase* . 107
Zts. physiol. Chem., **145**, 69.

R. EGE (1925), *Einfluss der Temperatur und der Reaktion auf Pepsin-destruktion und -aktivitat* 25, 66
Zts. physiol. Chem., **143**, 159.

W. ENGELHARDT (1924), *Über die Wirkung der Antiphenolase in adsorbiertem Zustande* 165
Bioch. Zts., **148**, 463.

H. ERDTMANN (1927), *Glycerophosphatspaltung durch Nierenphosphatase und ihre Aktivierung* 133
Zts. physiol. Chem., **172**, 182.

H. ERDTMANN (1928, 1), *Über Nierenphosphatase und ihre Aktivierung, II.* 133
Zts. physiol. Chem., **177**, 211.

H. ERDTMANN (1928, 2), *Über Nierenphosphatase, III.* 133
Zts. physiol. Chem., **177**, 231.

ERNSTRÖM (1922), *Über den Temperaturkoeffizienten der Stärkespaltung und die Thermostabilität der Malzamylase und des Ptyalins* . . 25
Zts. physiol. Chem., **119**, 190.

H. v. EULER and R. BLIX (1919), *Verstürkung der Katalasewirkung in Hefezellen* 71
Zts. physiol. Chem., **105**, 83.

H. v. EULER and J. BOLIN (1908), *Zur Kenntnis biologisch wichtiger Oxydationen, I.* 190, 192
Zts. physiol. Chem., **57**, 80.

H. v. EULER and J. BOLIN (1909, 1), *Zur Kenntnis biologisch wichtiger Oxydationen, II.* 190, 192
Zts. physiol. Chem., **61**, 1.

H. v. EULER and J. BOLIN (1909, 2), *Zur Kenntnis biologisch wichtiger Oxydationen, III.* 190, 192
Zts. physiol. Chem., **61**, 72.

H. v. EULER and E. BORGENSTAM (1920), *Zur Kenntnis der Katalasewirkung der Erythrozyten* 71
Bioch. Zts. **102**, 124.

H. v. EULER and E. BRUNIUS (1927), *Zur Kenntnis der Nukleosidasen, I.* . 63
Ber., **60**, 1584.

H. v. EULER and ERICSON (1922), *Neue Versuche über den Dispersitätsgrad der Saccharase* 179
Kolloid Zts., **31**, 3.

H. v. EULER and S. E. ERIKSSON (1926), *Zur Kenntnis der enzymatischen Spaltung des Sinigrins* 16
Fermentforschung, **8**, 518.

H. v. EULER and S. HEINTZE (1919), *Über die pH-Empfindlichkeit der Gärung einer Oberhefe* 18
Zts. physiol. Chem., **108**, 165.

H. v. EULER and K. JOSEPHSON (1923, 1), *Saccharase* 166
Ber. Deutsch. Chem. Ges., **56**, 446.

H. v. EULER and K. JOSEPHSON (1923, 2), *Inaktivierung der Saccharase durch Halogen* 152
Zts. physiol. Chem., **127**, 99.

H. v. EULER and K. JOSEPHSON (1924, 1), *Über die Affinität der Saccharase zu verschiedenen Zuckerarten* 48
Zts. physiol. Chem., **132**, 304.

H. v. EULER and K. JOSEPHSON (1924, 2), *Enzymatische Gleichgewichte, I.* . 82
Zts. physiol. Chem., **136**, 30.

H. v. EULER and K. JOSEPHSON (1924), *Versuche zur enzymatischen Spaltung der Saccharase, I.* 162, 167
Zts. physiol. Chem., **138**, 11.

H. v. EULER and K. JOSEPHSON (1924), *Versuche zur enzymatischen Spaltung der Saccharase, II.* 162, 167
Zts. physiol. Chem., **138**, 38.

Page of Text on which reference is made.

H. v. Euler and K. Josephson (1926), *Enzymatische Spaltung von Dipeptiden* 36, 63, 156
Zts. physiol. Chem., **157**, 122.

H. v. Euler and K. Josephson (1927, 1), *Katalase, I.* 1, 37, 166
Ann. Chem., **452**, 158.

H. v. Euler and K. Josephson (1927, 2), *Katalase, II.* 42
Ann. Chem., **455**, 1.

H. v. Euler and K. Josephson (1927, 3), *Katalase, III.* 5, 42
Ann. Chem., **455**, 111.

H. v. Euler and K. Josephson (1927, 4), *Enzymatische Spaltung von Dipeptiden. IV. Über die Wirkungsweise des Darmerepsins* . . 114
Zts. physiol. Chem., **162**, 85.

H. v. Euler, K. Josephson, and K. Myrbäck (1924), *Zur Berechnung der Aktivitäts -pH-Kurve der Saccharase* 21
Zts. physiol. Chem., **134**, 39.

H. v. Euler and I. Laurin (1919), *Über die Temperaturempfindlichkeit der Saccharase (Invertase)* 26, 69
Zts. physiol. Chem., **108**, 64.

H. v. Euler and I. Laurin (1920), *Über die Temperaturkoeffizienten der Saccharasewirkung.* 40, 67
Zts. physiol. Chem., **110**, 55.

H. v. Euler and K. Myrbäck (1923, 1) 151
Zts. exp. Med., **33**, 483, quoted by Myrbäck (1926, 1).

H. v. Euler and K. Myrbäck (1923, 2), *Garungs-Co-Enzym (Co-Zymase) der Hefe, I.* 37, 136
Zts. physiol. Chem., **131**, 179.

H. v. Euler and K. Myrbäck (1924), *Garungs-Co-Enzym (Co-Zymase der Hefe, III.* 136
Zts. physiol. Chem., **136**, 107.

H. v. Euler and K. Myrbäck (1927), *Zur Kenntnis der Enzymatischen Umwandlungen der Aldehyde, III.* 137
Zts. physiol. Chem., **165**, 28.

H. v. Euler and K. Myrbäck (1928), *Cozymase, XV.* 136, 137
Zts. physiol. Chem., **177**, 237.

H. v. Euler and K. Myrbäck (1929), *Cozymase. XVI. Weitere Isolierungsversuche* 136
Zts. physiol. Chem., **184**, 163.

H. v. Euler, K. Myrbäck, and R. Nilsson (1927), *Cozymase. XII. Das Molekulargewicht der Cozymase* 137
Zts. physiol. Chem., **168**, 177.

H. v. Euler, R. Nilsson, and B. Jansson (1927), *Cozymase, X.* . . . 137
Zts. physiol. Chem. **163**, 202.

H. v. Euler and F. Nordlund (1921), *Über die enzymatische Synthese des Fructose-phosphates* 18
Zts. physiol. Chem., **116**, 229.

H. v. Euler and O. Svanberg (1919), *Zur Kenntnis der Pektase-Wirkung* . 15
Biochem. Zts., **100**, 271.

H. v. Euler and O. Svanberg (1920), *Über Giftwirkungen bei Enzymreaktionen. I. Inaktivierung der Saccharase durch Schwermetalle* . 149
Fermentforschung, **3**, 330.

K. Fajans (1910), *Über die stereochemische Spezifität der Katalysatoren* . 5
Zts. physikal. Chem., 73, 25.

K. G. Falk (1915), *An experimental study of lipolytic actions* . . . 134
Proc. Nat. Ac. Sc., **1**, 136.

Famulener and Madsen (1908), *Die Abschwächung der Antigene durch Erwärmung* 69
Bioch. Zts., **11**, 186.

Page of Text on which reference is made.

G. Fauré (1835), *Nouvelles observations sur les semences de moutarde noire* 8
Journ. Pharm., **21**, 464.

F. Fenger (1923), *A comparison between the chemical and physiological properties of pepsin and rennin* 166
J. Amer. Chem. Soc., **45**, 249.

E. Fischer (1894), *Einfluss der Configuration auf die Wirkung der Enzyme, II.* 9, 98
Ber., **27**, 3479.

E. Fischer (1895), *Einfluss der Configuration auf die Wirkung der Enzyme* 98
Ber., **28**, 1429.

E. Fischer (1919), *Einfluss der Struktur der β-Glukoside auf die Wirkung des Emulsins* 96, 97, 98
Zts. physiol. Chem., **107**, 176.

E. Fischer and E. F. Armstrong (1902), *Synthese einiger neuer Disaccharide* 95
Ber., **35**, 3144.

E. Fischer and G. O. Curme (1914), *Über Lactal und Hydro-lactal* . . 95
Ber., **47**, 2047.

E. Fischer and W. Niebel (1896), *Über das Verhalten der Polysaccharide gegen einige thierische Sekrete und Organe* 100
Stz. ber. kgl. Pr. Ak. Wiss., **5**, 73.

E. Fischer and H. Thierfelder (1894), *Verhalten der verschiedenen Zucker gegen reine Hefen* 121
Ber., **27**, 3031.

E. Fischer and G. Zemplén (1910), *Verhalten der Cellobiose gegen einige Enzyme* 95
Ann. Chem., **372**, 254.

A. Fleisch (1924), *Some oxidation processes of normal and cancer tissue* . 156
Bioch. J., **18**, 294.

P. Fleury (1925), *La laccase et les lois de l'action des diastases* 18
J. pharm. et chim., **1**, 105.

A. Fodor (1926), *Fermentwirkung und Wasserstoffionenkonzentration* . . 176
Koll. Ztsch., **40**, 234.

J. C. Forbes (1927), *The purification of pepsin, its properties and physical characters* 170
J. Biol. Chem., **71**, 559.

R. Fosse and A. Brunel (1929), *Un nouveau ferment* 188
C. R. Ac. Sc., **188**, 426.

D. L. Foster (1925), *The relation between the pancreas and carbohydrate metabolism of muscle. II. Antiglyoxalase and glyoxalase* 163
Bioch. J., **19**, 757.

Fränkel and Jellinek (1927), *Über die sogenannte Kohlehydratgruppe im Eiweiss (Darstellung der Glucosamino-mannose)* 116
Bioch. Zts., **185**, 392.

H. Freundlich (1922), *Capillarchemie*, p. 232 42

F. W. Geddes and A. Hunter (1928), *Observations upon the enzyme asparaginase* 17, 107
J. Biol. Chem., **77**, 197.

Goldschmidt (1925), in Oppenheimer : " *Die Fermente* " (5th ed.), p. 235 91

Gottschalk (1926), *Aufbau und Vergärung von Glykogen durch maltasefrei Hefe* 117
Zts. physiol. Chem., **152**, 132.

*W. Grassmann (1927), *Über die Dipeptidase und die Polypeptidase der Hefe* 17
Zts. physiol. Chem., **167**, 202.

W. Grassmann (1928), *Neue Methoden und Ergebnisse der Enzymforschung* 172
(Bergmann).

Page of Text on which reference is made.

*W. GRASSMANN and H. DYCKERHOFF (1928, 1), *Über die Spezifität der Hefe-Peptidase* 112, 113
Ber., **61**, 656.

*W. GRASSMANN and H. DYCKERHOFF (1928, 2), *Über die Wirkungsweise der Hefepolypeptidase* 108, 113
Zts. physiol. Chem., **175**, 18.

*W. GRASSMANN and H. DYCKERHOFF (1928, 3), *Über die Proteinase und die Polypeptidase der Hefe* 23, 113, 144, 156
Zts. physiol. Chem., **179**, 41.

W. GRASSMANN, H. DYCKERHOFF, and O. v. SCHOENEBECK (1930), *Über naturliche Aktivatoren und Hemmungskorper proteolytischer Enzyme* 108, 142, 144, 156, 157
Zts. physiol. Chem., **186**, 183.

W. GRASSMANN and L. KLENK (1930), *Zur Frage der Einheitlickkeit tierischen und pflanzlichen Dipeptidase* 109
Zts. physiol. Chem., **186**, 26.

WL. GULEWITSCH (1899), *Über das Verhalten des Trypsins gegen einfachere chemische Verbindungen* 114
Zts. physiol. Chem., **27**, 540.

E. HÄGGLUND and A. RINGBOM (1927), *Über die Vergärung der α-Ketobuttersäure und Oxalessigsäure. VII. Über die Abhängigkeit von der Wasserstoffionen-Konzentration* 18
Bioch. Zts., **187**, 117.

E. HÄGGLUND and T. ROSENQUIST (1926), *Über die Abhängigkeit der alkoholischen Gärung von der Wasserstoffionen Konzentration* . . 18
Bioch. Zts., **175**, 293.

E. HÄGGLUND and T. ROSENQUIST (1927), *Zur Kenntnis der Kinetik der Carboxylase-Wirkung* 37
Bioch. Zts., **181**, 296.

D. E. HALEY and J. F. LYMAN (1921), *Castor bean lipase. Its preparation and some of its properties* 15
J. Amer. Chem. Soc., **43**, 2664.

A. HARDEN (1923), "*Alcoholic Fermentation.*" (This series) 4, 189

A. HARDEN and YOUNG (1906), *The alcoholic fermentation of yeast-juice. Part II. The co-ferment of yeast-juice* 133, 136
Proc. Roy. Soc., **77**B, 405.

M. L. C. HARE (1928), *Tyramine oxidase* 11, 18
Bioch. J., **22**, 968.

L. J. HARRIS (1923), *On the existence of an unidentified sulphur grouping in the protein molecule. I. On the denaturation of proteins* . . . 116
Proc. Roy. Soc., **94**B, 426.

D. C. HARRISON (1927), *The catalytic action of traces of iron and copper on the anaerobic oxidation of sulphydryl compounds* 157
Bioch. J., **21**, 335.

Y. HATTORI (1924), *Das Verhalten von α- und β-methylglucosid zur Takainvertase* 16, 43
Bioch. Zts., **150**, 150.

*F. HAUROWITZ and W. PETROU (1924), *Über das pH Optimum der Magenlipase verschiedener Tiere* 15
Zts. physiol. Chem., **144**, 68.

W. N. HAWORTH (1927), *Structural relationships in the carbohydrate group* . 117
J. Soc. Chem. Industry, **46**, 299T.

W. N. HAWORTH (1929), *The constitution of sugars* (Arnold) 131

J. J. HAZEWINKEL (1900), *Indican, its hydrolysis and the enzyme causing the same* 188
Kon. Akad. Wet. Amsterdam, Proceedings of the section of sciences, **2**, 512.

Page of Text on which reference is made.

S. HEDIN (1907), *A case of specific adsorption of enzymes* 30
Bioch. J. **2**, 112.

B. HELFERICH, W. KLEIN, and W. SCHÄFER (1926), *Zur Spezifität der α-Glucosidase aus Hefe* 97
Ber., **59**, 79.

B. HELFERICH and J. BECKER (1924), *Synthese einer Disaccharidglucosids*. 97, 98
Ann. Chem., **440**, 1.

B. HELFERICH, A. LOEWA, W. NIPPE, and H. RIEDEL (1923), *Über die Einwirkung von Fermenten auf Schwefelsaüre- und Phosphorsaüreester der Zucker und ihrer Derivate* 96, 98
Zts. physiol. Chem., **128**, 141.

B. HELFERICH, W. KLEIN, and W. SCHAFER (1926), *Zur Spezifität der α-Glukosidase aus Hefe* 97
Ber. **59**, 79.

B. HELFERICH and H. RAUCH (1927), *Synthese von Primeverose*. . . . 131
Ann. Chem., **455**, 168.

B. HELFERICH, P. E. SPEIDEL, and W. TOELDTE (1923), *Über Emulsin* . 162
Zts. physiol. Chem., **128**, 99.

V. HENRI (1903), " *Lois generales de l'action des diastases* " 9, 38

V. HENRI (1906), *Action de l'invertine dans un milieu hétérogène* . . 6
C. R. Ac. Sc., **142**, 97.

V. HENRI and CH. PHILOCHE (1904), *Ralentissement de l'action de la maltase par le glucose et par le lévulose* 63
C. R. Soc. Biol., **2**, 170.

HERZOG (1913), in Oppenheimer: " *Die Fermente* " (4th ed.), p. 1021 . 91

C. N. HINSHELWOOD (1926), " *The Kinetics of Chemical Change in Gaseous Systems* " (Oxford), p. 12 177

H. F. HOLDEN (1924), *Experiments on respiration and fermentation* . . 137
Bioch. J., **18**, 535.

F. G. HOPKINS (1925), *Glutathione. Its influence on the oxidation of fats and proteins* 145
Bioch. J., **19**, 787.

C. S. HUDSON (1908), *Inversion of sucrose by invertase, I.* 11
J. Amer. Chem. Soc., **30**, 1160.

C. S. HUDSON (1909), *Inversion of sucrose by invertase, III*. 85
J. Amer Chem. Soc., **31**, 655.

C. S. HUDSON and H. S. PAINE (1910), *The inversion of cane sugar by invertase. VII. The effect of alcohol on invertase* 160
J. Amer. Chem. Soc., **32**, 1350.

L. HUGOUNENQ and J. LOISELEUR (1925), *Sur la constitution des diastases protéolytiques et le mécanisme de leur action* 23
Bull. Soc. Chim. Biol., **7**, 955.

A. HUNTER and J. DAUPHINEE (1924), *An approximative colorimetric method for the determination of urea, with an application to the detection and quantitative estimation of arginase* 17
Proc. Roy. Soc., **97**B, 209.

R. G. HUSSEY and J. H. NORTHROP (1923), *A study of the equilibrium between the so-called antitrypsin of the blood and trypsin* 163
J. gen. physiol., **5**, 335.

R. G. HUSSEY and W. R. THOMPSON. *The effect of radio-active emanation and X-rays on enzymes* 73
(1923, 1), I., J. gen. physiol., **5**, 647.
(1923, 2), II., J. gen. physiol., **6**, 1.
(1925) IV., J. gen. physiol., **9**, 211.
(1926, 1), V., J. gen. physiol., **9**, 309.
(1926, 2), VI., J. gen. physiol., **9**, 315.

Page of Text on which reference is made.

T. Imai (1924), *Über die methylierten Eiweisstoffe. I. Über die Wirkung der proteolytischen Enzyme auf die methylierten Eiweisstoffe* . . . 23
Zts. physiol. Chem., **136**, 173.

W. Issajew (1904), *Über die Hefekatalase* 37
Zts. physiol. Chem., **42**, 102.

R. Iwatsuru (1926), *Über die Spaltung der mono-phenyl-phosphorsäuren und mono-äthyl-phosphorsäuren Salze durch pflanzliche und tierische Phosphatase* 105
Bioch. Zts., **173**, 348.

K. Jamada and A. Jodlbauer (1908), *Die Wirkung des Lichtes auf Peroxydase und ihre Sensibilisierung durch fluoreszierende Stoffe* . . 72, 73
Bioch. Zts., **61**, 8.

F. Johannesson (1917), *Der Einfluss des Formaldehyde auf die Eiweissverdauung* 23
Bioch. Zts., **83**, 28.

W. Jones (1914), "*Nucleic Acids.*" (This series), p. 78 188

K. Josephson (1924), *Über die Affinitat der Saccharase zu verschiedenen Zuckern, II.* 14, 39, 48, 49
Zts. physiol. Chem., **134**, 50.

K. Josephson (1925, 1), *Enzymatische Spaltung von Glucosiden* 16, 33, 36, 48, 49, 82
Zts. physiol. Chem., **147**, 1.

K. Josephson (1925, 2), *Enzymatische Spaltung von Glucosiden*. . 36, 49, 82
Zts. physiol. Chem., **147**, 155.

K. Josephson (1926), *Die Enzyme des Emulsins, II.* 166
Ber. Deutsch. Chem. Ges., **59**, 821.

H. Jost (1927), *Über die Bedeutung der Saüreloslichen organischen Phosphors* 106
Zts. physiol. Chem., **165**, 171.

Karrer (1926), "*Tschirsch—Festschrift.*" (Quoted by Erdtmann (1927)). 15
Zts. physiol. Chem., **172**, 183.

P. Karrer and A. Hofmann (1929), *Über den enzymatischen Abbau von Chitin und Chitosan* 17
Helv. Chim. Act., **12**, 616.

J. H. Kastle and A. S. Loevenhart (1900), *Lipase, the fatsplitting enzyme, and the reversibility of its action* 65, 66
J. Am. Chem. Soc., **24**, 491.

J. H. Kastle (1906), *Hyg. Lab. Treas. Bull.*, 26. (Quoted by Euler: "*Chemie der Enzyme,*" II., 1, p. 28) 152

N. Kato, (1923, 1) *Über die Beziehung zwischen Harnstoffkonzentration und Ureasewirkung, und den Einfluss des Glykokolls auf dieselbe* . 146
Bioch. Zts., **136**, 498.

N. Kato (1923, 2), *Über den stabilen Bestandteil des Soja-Urease-Präparats*. 146, 147
Bioch. Zts., **139**, 352.

H. D. Kay (1923), *The reversibility of the action of urease of soy bean* . . 2
Bioch. J., **17**, 277.

H. D. Kay (1926), *Kidney phosphatase* 106
Bioch. J., **20**, 791.

H. D. Kay (1928, 1), *The phosphatase of mammalian tissues, I.* . . . 15, 105
Bioch. J., **22**, 855.

H. D. Kay (1928, 2), *The phosphatase of mammalian tissue. II. Pyrophosphatase* 15, 105
Bioch. J., **22**, 1446.

D. Keilin (1925), *On cytochrome, a respiratory pigment, common to animals, yeast, and higher plants* 174
Proc. Roy. Soc., **98**B, 312.

D. Keilin (1929), *Cytochrome and respiratory enzymes* 157, 191, 192
Proc. Roy. Soc., **104**B, 206.

Page of Text on which reference is made.

E. C. KENDALL and H. C. SHERMAN (1910), *Amylase. II. Action of pancreatic amylase* 36
J. Amer. Chem. Soc., **32**, 1087.

KIESEL (1922), *Über die Wirkung der Arginase auf Agmatin und Tetramethylendiguanidin* 107
Zts. physiol. Chem., **118**, 284.

H. KIMURA (1929), *Studien über die fermentative Spaltung der Acetyl-Formyl- und Benzoylderivate der Aminosaüren* 17, 107
J. Bioch., **10**, 207.

H. KIMURA (1929), *Weitere Studien über die fermentative Spaltung der Acetyl- Formyl- und Benzoylderivate der Aminosaüren*. 107
J. Bioch., **10**, 225.

T. KITASATO (1927), *Partielle Hydrolyse des Populins zu Saligenin und Benzoylglucose durch ein Enzym der Takadiastase* 98
Bioch. Zts., **190**, 109.

T. KITASATO (1928), *Über Meta-Phosphatase* 105
Bioch. Zts., **197**, 257.

A. J. KLUYVER and A. P. STRUYK (1927), *Die Existenz der Zellfreien Garung* 8
Zts. physiol. Chem., **170**, 110.

F. KNAFFL-LENZ (1923), *Über die Kinetik der Esterspaltung durch Leberlipase* 15
Arch. exp. Path. Pharm., **97**, 242.

H. KOBAYASHI (1926), *Über die Glycerophosphatase* 35
J. Bioch., **6**, 261.

H. KOBAYASHI (1927), *Über die Glycerophosphatase* 15
J. Bioch., **8**, 205.

H. KOBAYASHI (1929), *Die fermentative Spaltung der Diphosphoglycerinsaüre* 15
J. Bioch., **11**, 173.

A. KOSSEL and H. D. DAKIN (1906), *Weitere Untersuchungen über fermentative Harnstoffbildung* 107
Zts. physiol. Chem., **42**, 181.

S. KOSTYTSCHEW, G. MEDWEDEW, H. KARDO-SYSOJEWA (1927), *Über Alkoholgarung. XIII. Die Nichtexistenz der Zellfreien Garung* . . 8
Zts. physiol. Chem., **168**, 244.

H. KRAUT (1928), *Methoden der Adsorption und Elution.* (Oppenheimer und Pincussen, "*Die Methodik der Fermente*" (1928), p. 445) . 170, 171, 172

*H. KRAUT and H. RUBENBAUER (1928), *Über Leberesterase* . . . 166, 175
Zts. physiol. Chem., **173**, 103.

H. A. KREBS (1928), *Über die Wirkung von CO und Licht auf Häminkatalysen* 190
Bioch. Zts., **193**, 347.

V. K. KRIEBLE (1913), *Enzymes. Asymmetric syntheses through the action of oxynitrilases, I.* 128
J. Amer. Chem. Soc., **35**, 1643.

V. K. KRIEBLE and W. A. WIELAND (1921), *The properties of oxynitrilase* . 128
J. Amer. Chem. Soc., **43**, 164.

*R. KUHN (1923, 1), *Über Spezifität der Enzyme. II. Saccharase- und Raffinasewirkung des Invertins* 31, 32, 36, 41, 43, 45
Zts. physiol. Chem., **125**, 28.

*R. KUHN (1923, 2), *Über Spezifität der Enzyme. III. Die Affinität der Enzyme zu stereoisomeren Zuckern* 48
Zts. physiol. Chem., **127**, 234.

*R. KUHN (1924), *Über den Einfluss stereoisomerer Zucker sowie nicht spaltbarer Kohlehydrate und Glucoside auf die Wirksamkeit des Hefe-Invertins.* 48
Zts. physiol. Chem., **135**, 1.

R. KUHN (1925, 1), *Über die Wirkungsweise der Amylase* 117, 118
Ann. Chem., **444**, 1.

R. KUHN (1925 2), Oppenheimer's "*Die Fermente,*" p. 240 91

Page of Text on which reference is made.

R. Kuhn and L. Brann (1926), *Über die Abhängigkeit der Katalatischen und peroxydatischen Wirkung des Eisens von seiner Bindungsweise* . 21
Ber., **59**, 2370.

R. Kuhn and G. Grundherr (1926), *Die Konstitution der Melezitose und Turanose* 100
Ber. Deutsch. Chem. Ges., **59**, 1655.

R. Kuhn and R. Heckscher (1926), *Über die enzymatische Bildung von Milchsaüre aus Methylglyoxal* 18, 36, 37, 163
Zts. physiol. Chem., **160**, 116.

R. Kuhn and H. Münch (1925), *Über Gluco- und Fructosaccharase, I.* 99, 100, 101
Zts. physiol. Chem., **150**, 220.

R. Kuhn and H. Münch (1927), *Über Gluco- und Fructosaccharase, II.* 47, 48, 49
Zts. physiol. Chem., **163**, 1.

R. Kuhn and H. Schlubach (1925), *Über das Verhalten permethylierter Zucker gegen Emulsin* 98
Zts. physiol. chem., **143**, 154.

R. Kuhn and T. Wagner-Jauregg (1927), *Über die Einwirkung von Enzymen auf das γ-Methylglucosid von Emil Fischer* 120
Zts. physiol. Chem., **162**, 103.

R. Kuhn and W. Ziese (1926), *Über die Verknupfungsstelle der Traubenzuckerreste in der Stärke. Abbau von Monomethyl-trihexosan zur* 6-*Methyl-Glucose* 116
Ber., **59**, 2314.

S. Kumanomidoh (1928), *Diastase, V.* 71
Bioch. Zts., **195**, 79.

I. Langmuir (1916), *The constitution and fundamental properties of solids and liquids. Part I. Solids* 28, 58
J. Amer. Chem. Soc., **38**, 2221.

Lebedew and Griaznow (1912), *Über der Mechanismus der Alkoholischen Gärung* 120
Ber., **45**, 3256.

J. Leibowitz (1925), *Über die Gerstenmalzmaltase und die Spezifität der Disaccharasen* 95
Zts. physiol. Chem., **149**, 184.

J. Leibowitz and P. Mechlinski (1926), *Über Takamaltase und Takasaccharase* 95
Zts. physiol. Chem., **154**, 64.

J. Leibowitz and P. Mechlinski (1926), *Über die fermentative Spaltung der Polyamylosen* 16, 116
Ber., **59**, 2738

Leuchs (1831), " Kestner's Archiv fur die gesamte Naturlehre." (Quoted by Oppenheimer in " *Die Fermente und Ihre Wirkungen,*" p. 717) . 8

P. A. Levene and T. Mori (1929), *The carbohydrate group of ovomucoid* . 116
J. Biol. Chem., **84**, 49.

P. A. Levene, W. Jacobs and F. Medigreceanu (1912), *On the action of tissue extracts containing nucleosidase on α- and β-methylpentosides* . 187
J. Biol. Chem., **11**, 371.

P. A. Levene, H. S. Simms, and M. H. Pfaltz (1926), *Enzyme hydrolysis of dipeptides and tripeptides* 110
J. Biol. Chem., **70**, 253.

P. A. Levene, R. E. Steiger, and L. W. Bass (1929), *The relation of chemical structure to the rate of hydrolysis of peptides. V. Enzyme hydrolysis of dipeptides* 110, 114
J. Biol. Chem.

Page of Text on which reference is made.

P. A. LEVENE, M. YAMAGAWA, and J. WEBER (1924), *On nucleosidases. I. General properties* 15, 35
J. Biol. Chem., **60,** 693.

P. C. LEWIS (1926, 1), *The effect of variation in the hydrogen ion concentration on the velocity of the heat denaturation of ox hæmoglobin* 69
Bioch. J., **20,** 965.

P. C. LEWIS (1926, 2), *The effect of variation in the hydrogen ion concentration on the velocity of the heat denaturation of egg albumin. The critical increment of the process* 69
Bioch. J., **20,** 978.

J. v. LIEBIG (1870), *Chemische Briefe*, 21 8

J. v. LIEBIG and WÖHLER (1837), *Üeber die Bildung des Bittermandelöls* . 8
Ann. Chem., **22,** 1.

K. H. LIN, H. WU, T. T. CHEN (1927), *Effect of racemisation on digestibility of casein and egg albumin by pepsin and trypsin* 111
Proc. Roy. Soc. Exp. Biol. Med., **25,** 201.

K. H. LIN, H. WU, T. T. CHEN (1928), *Digestibility of racemised casein and egg albumin* 111
Chinese J. Physiol. **2,** 131.

K. LINDERSTRØM-LANG and M. SATO (1929), *Die Spaltung von Glyzylglyzin, Alanylalanin und Leuzylglycin durch Darm- und Malzpeptidasen* 17, 24, 109
Zts. physiol. Chem., **184,** 83.

A. R. LING and D. R. NANJI (1923), *Studies on starch. I. The nature of polymerised amylose and of amylopectin* 117
J. Chem. Soc., **123,** 2666.

A. R. LING and D. R. NANJI (1925), *Studies on starch. Part III.* . . . 117
Trans. Chem. Soc., **127,** 636.

C. J. LINTNER and E. KRÖBER (1895), *Zur Kentniss des Hefeglycase* . 66
Ber. Deutsch. Chem. Ges., **28,** 1050.

G. LJUNGGREN (1925), *Katalytisk Kolesyreavspjälkning us ketokarbonsyrer* . 4
(Lund University Press.)

O. LOEW (1901), *Catalase, a new enzyme of general occurrence* 155
Rep. 68, U.S. Dept. Agric. (cited by Oppenheimer, *Die Fermente*).

O. LOEWI and E. NAVRATIL (1926), *Über das Schicksal des Vagusstoffens und des Acetylcholins im Herzen* 102
Klin. Woch., **5,** 894.

K. LOHMANN (1926), *Über die Hydrolyse des Glykogens durch die diastatischen Fermente des Muskels* 16, 36, 117, 119
Bioch. Zts., **178,** 444.

K. LOHMANN (1928), *Über das Vorkommen und den Umsatz von Pyrophosphat in Zellen, III.* 105
Bioch. Zts., **203,** 172.

C. LOVAT EVANS (1912), *The amyloclastic property of saliva* 34, 36
J. Physiol. **44,** 191.

S. LÖVGREN (1921), *Studien über die Urease* 14, 17
Bioch. Zts., **119,** 285.

S. LÖVGREN (1927), Cited by Euler, "*Chemie der Enzyme,*" *II.*, p. 346 . 106

H. LUERS and F. ALBRECHT (1926), *Über Antiamylase. Ein Beitrag zur Frage der Antienzyme* 164
Fermentforschung, **8,** 52.

H. LUERS and W. WASMUND (1922), *Über die Wirkungsweise der Amylase* . 66, 69
Fermentforschung, **5,** 169.

E. MACK and D. S. VILLARS (1923), *The action of urease in the decomposition of urea* 2
J. Amer. Chem. Soc., **45,** 505.

H. R. MARSTON (1923), *The azine and azonium compounds of the proteolytic enzymes, I.* 170
Bioch. J., **17,** 850.

Page of Text on which reference is made.

M. MARTLAND and R. ROBINSON (1927), *Bone phosphatase. VII. The possible significance of hexose esters in ossification* 35, 63, 105
Bioch. J., **21**, 665.

MAYER (1906), *Über die Spaltung der Lipoiden* 103
Bioch. Zts., **1**, 39.

P. MAYER (1913), *Zuckerfreie Garung bei Stereoisomeren* 122
Bioch. Zts., 50, 283.

P. MAYER (1928), *Vergleichende Untersuchungen über die Wirkung von Arsenat und von organischen Arsensaüre-abkommlingen auf die alkoholische Zuckerspaltung* 134
Bioch. Zts., **193**, 176.

R. A. McCANCE (1925, 2), *Tyrosinase, its action on phenols, tyrosine, and other amino-acids* 191
Bioch. J., **19**, 1022.

R. A. McCANCE (1925, 1), *The influence of oxygen on the production of urea by enzymes of the liver and spleen* 153
Bioch. J., **19**, 134.

J. MELLANBY and V. WOOLLEY (1913), *The ferments of the pancreas. Part I. The generation of trypsin from trypsinogen by enterokinase* . . . 141
Journ. Physiol., **45**, 370.

K. MEYER (1928), *Über die Reinigung des Milchsaürebildenden Ferments* . 160, 166
Bioch. Zts., **193**, 139.

O. MEYERHOF (1914), *Die Hemmung von Fermentreaktionen durch indifferente Narkotika* 158
Pflüg. Arch., **157**, 251.

O. MEYERHOF (1918, 1), *Über das Vorkommen des Coferments der alkolischen Hefegärung in Muskelgewebe und seine mutmassliche Bedeutung im Atmungsmechanismus* 137, 145
Zts. Phys. Chem., **101**, 165.

O. MEYERHOF (1918, 2), *Untersuchungen zur Atmung getöteter Zellen. II. Der Oxydationsvorgang in getöteter Hefe und Hefe-extrakt* 137, 145
Pflüg. Arch., **170**, 367.

O. MEYERHOF (1926), *Über die enzymatische Milchsaürebildung im Muskelextrakt. II. Mitteilung. Die Spaltung der Polysaccharide und der Hexosediphosphorsaüre* 37
Bioch. Zts., **178**, 462.

O. MEYERHOF (1927), *Über die enzymatische Milchsaürebildung im Muskelextrakt. III. Die Milchsaürebildung aus den gärfahigen Hexosen* 134, 137, 144
Bioch. Zts., **183**, 176.

L. MICHAELIS (1908), *Die Adsorptionsaffinitäten des Hefe-Invertins* . . 28
Bioch. Zts., **7**, 488.

L. MICHAELIS (1921), *Weitere Beiträge zur Theorie der Invertasewirkung* . 29
Bioch. Zts., **115**, 269.

L. MICHAELIS and H. DAVIDSOHN (1910), *Die Isoelektrische Konstante des Pepsins* 22
Bioch. Zts., **28**, 1.

L. MICHAELIS and H. DAVIDSOHN (1911), *Die Wirkung der Wasserstoffionen auf das Invertin* 19, 20
Bioch. Zts., **35**, 386.

L. MICHAELIS and M. EHRENREICH (1908), *Die Adsorptionsanalyse der Fermente* 170
Bioch. Zts., **10**, 283.

L. MICHAELIS and MENDELSOHN (1914), *Die Wirkungsbedingung des Labferments* 17
Bioch. Zts., **58**, 315.

L. MICHAELIS and M. L. MENTEN (1913), *Die Kinetik der Invertinwirkung* . 38, 75
Bioch. Zts., **49**, 333.

L. MICHAELIS and H. PECHSTEIN (1913), *Untersuchungen über die Katalase der Leber* 18, 20
Bioch. Zts., **53**, 320.

Page of Text on which reference is made.

L. MICHAELIS and H. PECHSTEIN (1914, 1), *Die Wirkungsbedingungen der Speicheldiastase* 138, 151
Bioch. Zts., **59**, 77.

L. MICHAELIS and H. PECHSTEIN (1914, 2), *Über die verschiedenartige Natur der Hemmungen der Invertasewirkung* 46
Bioch. Zts., **60**, 79.

L. MICHAELIS and P. RONA (1913), *Die Wirkungsbedingungen der Maltase von Bierhefe, I.* 15, 20
Bioch. Zts., **57**, 70.

K. MIYAKE and M. ITO (1924), *On lethal temperature of pure Koji diastase in aqueous solution and recovery of its action after heating* . . . 69, 70
J. Bioch., **3**, 177.

W. MORGAN and R. ROBISON (1928), *Constitution of hexose-diphosphoric acid. Part II. The dephosphorylated α- and β-methylhexosides* . . 131
Bioch. J., **22**, 1270.

S. MORGULIS and N. BEBER (1928), *Studies on the effect of temperature on the catalase reaction. VI. Heat inactivation of catalase at different hydrogen ion concentrations* 79
Journ. Biol. Chem., **77**, 115.

D. R. P. MURRAY (1929), *Molecular constitution and accessibility to enzymes. The effect of various substances on the velocity of hydrolysis by pancreatic lipase* 52
Bioch. J., **23**, 292.

D. R. P. MURRAY and C. G. KING (1930), *The stereochemical specificity of esterases, I.* 53
Bioch. J., **24**,, 190.

N. MUTCH (1912), *A short quantitative study of histozyme, a tissue ferment* . 63, 106
J. Physiol., **44**, 176.

K. MYRBÄCK (1926, 1), *Über Verbindungen einiger Enzyme mit inaktivierenden Stoffen, I.* 149-155
Zts. physiol. Chem., **158**, 160.

K. MYRBÄCK (1926, 2), *Über Verbindungen einiger Enzyme mit inaktivierenden Stoffen, II.* 16, 17, 36, 138, 149-155
Zts. physiol. Chem., **159**, 1.

K. MYRBÄCK and W. JACOBI (1926), *Zur Kenntnis der enzymatischen Umwandlungen der Aldehyde, II.* 18, 137
Zts. physiol. Chem., **161**, 245.

K. MYRBÄCK and R. NILSSON (1927), *Cozymase, XI.* 137
Zts. physiol. Chem., **165**, 140.

J. M. NELSON and R. ANDERSON (1926), *Glucose and fructose retardation of invertase action* 40, 47, 48
J. Biol. Chem., **69**, 443.

J. M. NELSON and G. BLOOMFIELD (1924), *Some characteristics of invertase action* 13, 14, 16, 21, 67, 68
J. Amer. Chem. Soc., **46**, 1025.

J. M. NELSON and O. BODANSKY (1925), *Mutarotation as a factor in the kinetics of invertase action* 45, 49, 76
J. Amer. Chem. Soc., **47**, 1624.

J. M. NELSON and D. J. COHN (1924), *Invertase in honey* 14, 16, 36, 51
J. Biol. Chem., **61**, 193.

J. M. NELSON and E. G. GRIFFIN (1916), *The adsorption of invertase* . . 28
J. Amer. Chem. Soc., **38**, 1109.

J. M. NELSON and D. J. HITCHCOCK (1921), *Uniformity in invertase action* . 76
J. Amer. Chem. Soc., **43**, 2632.

J. M. NELSON and D. P. MORGAN (1923), *Collodion membranes of high permeability* 172
J. Biol. Chem., **58**, 305.

Page of Text on which reference is made.

J. M. NELSON and M. P. SCHUBERT (1928), *Water concentration and the rate of hydrolysis of sucrose by invertase* 30, 32, 182
J. Amer. Chem. Soc., **50**, 2188.

J. M. NELSON and C. T. SOTTERY (1924), *Influence of glucose and fructose on the rate of hydrolysis of sucrose by invertase from honey* 51, 52
J. Biol. Chem., **62**, 139.

NENCKI (1886), *Über die Spaltung der Saüreester der Fettreihe und der aromatischen Verbindungen im Organismus und durch das Pankreas* . 102, 106
Arch. exp. Path., **20**, 367.

C. NEUBERG (1913), *Weitere Untersuchungen über die biochemische Umwandlung von Methylglyoxal in Milchsaüre nebst Bemerkungen über die Entstehung der verschiedenen Milchsaüren in der Natur* 127
Bioch. Zts., **51**, 485.

C. NEUBERG and G. GORR (1925, 1), *Über den Mechanismus der Milchsaürebildung bei Bakterien, I.* 127
Bioch. Zts., **162**, 490.

C. NEUBERG and G. GORR (1925, 2), *Über den Mechanismus der Milchsaürebildung bei Bakterien, II.* 127
Bioch. Zts., **166**, 482.

C. NEUBERG and K. P. JACOBSOHN (1928), *Fortgesetzte Untersuchungen über den Wirkungsbereich der Phosphatase* 105, 106
Bioch. Zts., **199**, 498.

C. NEUBERG and M. KOBEL (1927), *Über die enzymatische Abspaltung von Methylalkohol aus Pektin durch ein Ferment des Tabaks* 15
Bioch. Zts., **190**, 232.

C. NEUBERG and J. LEIBOWITZ (1927, 1), *Über die partielle De-phosphorylierung der Hexose-di-phosphorsaüre durch Hefe* 133
Bioch. Zts., **191**, 450.

C. NEUBERG and J. LEIBOWITZ (1927, 2), *Abbau von Zymo-di-phosphat mittels tierischer Phosphatase zu Hexose-mono-phosphorsaüre-ester* . 133
Bioch. Zts., **191**, 456.

C. NEUBERG and J. LEIBOWITZ (1927, 3), *Arsenat-aktivierung und die Spezifität von Phosphatase* 133
Bioch. Zts., **191**, 460.

C. NEUBERG and J. LEIBOWITZ (1928), *Biochemische Darstellung eines Disaccharid-mono-phosphorsaure-esters* 133
Bioch. Zts., **193**, 237.

C. NEUBERG and P. MAYER (1903), *Über das Verhalten stereoisomerer Substanzen im Tierkorper. II. Über das Schicksal der drei Monosen im Kaninchenleber* 2
Zts. physiol. Chem., **37**, 530.

C. NEUBERG and W. H. PETERSON (1914), *Die Valeraldehyd- und Amylalkoholgärung der Methyläthylbrenztraubenäsure* 122
Bioch. Zts., **67**, 32.

C. NEUBERG and H. POLLAK (1910), *Über Phosphorsaüre- und Schwefelsaüreester von Kohlehydraten* 99
Bioch. Zts., **26**, 514.

C. NEUBERG and ROSENBERG (1907), *Verwandlung von optisch-inaktiven Triolein in ein optisch-aktives Glycerid und eine optisch-aktive säure* . 104
Bioch. Zts., **7**, 191.

C. NEUBERG and S. SANEYOSHI (1911), *Über den Nachweis kleiner Mengen von Disacchariden* 95
Bioch. Zts., **36**, 44.

C. NEUBERG and J. WAGNER (1925), *Über die Spaltbarkeit der Äthylschwefelsäuren der verschiedenen Reihen durch Sulfatase und über den Mechanismus der Sulfatase-Wirkung* 106
Bioch. Zts., **161**, 492.

C. NEUBERG and J. WAGNER (1926, 1), *Zur Kenntnis der Phosphatase und über die Darstellung von sauren Estern der Pyrophosphorsäure* . . . 105
Bioch. Zts., **171**, 485.

Page of Text on which reference is made.

C. Neuberg and J. Wagner (1926, 2), *Über die Verschiedenheit der Sulfatase und Myrosinase* 106
Bioch. Zts., **174**, 457.

C. Neuberg, J. Wagner, and K. P. Jacobsohn (1927), *Über die asymmetrische Wirkungsweise der Phosphatase und eine Methode zur biochemischen Darstellung der beiden entgegengesetzt drehenden Alkohole aus ihren Racematen* 106
Bioch. Zts., **188**, 227.

C. Neuberg and F. Weinmann (1928), *Strukturchemische Spezifitat der Carboxylase* 122
Bioch. Zts., **200**, 472.

S. M. Neuschloss (1920), *Untersuchungen über den Einfluss der Neutralsalze auf die Fermentwirkung* 158
Pflüg. Arch., **181**, 45.

M. Nicloux (1904), *Étude de l'action du cytoplasma de la graine de ricin. Action de la température* 69
C. R. Soc. Biol., 839.

K. Nishikawa (1927), *Zur Kenntnis der Takadiastase* 17, 29
Bioch. Zts., **188**, 387.

E. Nordefeldt (1920), *Über den Temperaturkoeffizienten der H_2O_2-Spaltung durch Fettkatalase* 66
Bioch. Zts., **109**, 236.

E. Nordefeldt (1921), *Die Bedeutung der Azidität fur die Oxynitrilsynthese und die Nichtexistenz des Rosenthalerschen syn-Emulsins* . . 128
Bioch. Zts., **118**, 15.

E. Nordefeldt (1922), *Über die asymmetrische Wirkung des Emulsins bei der Benz-oxynitril-synthese* 128
Bioch. Zts., **131**, 390.

E. Nordefeldt (1923), *Über die Wirkung des Emulsins auf das System Blausäure-Benzaldehyd-Benzoxynitril* 128
Bioch. Zts., **137**, 489.

J. H. Northrop (1919), *The combination of enzyme and substrate, I., II.* . 29, 30
J. gen. physiol., **2**, 113.

J. H. Northrop (1920, 1), *The influence of hydrogen-ion concentration on the inactivation of pepsin solutions* 25
J. gen. physiol., **2**, 465.

J. H. Northrop (1920, 2), *The effect of the concentration of enzyme on the rate of digestion on proteins by pepsin* 11, 12, 54
J. gen. physiol., **2**, 471.

J. H. Northrop (1920, 3), *The influence of the substrate concentration on the rate of hydrolysis of proteins by pepsin* 36, 54
J. gen. physiol., **2**, 595.

J. H. Northrop (1922, 1), *The inactivation of trypsin, I.* 54, 55
J. gen. physiol., **4**, 227.

J. H. Northrop (1922, 2), *The inactivation of trypsin, II.* 54, 55
J. gen. physiol., **4**, 245.

J. H. Northrop (1922, 3), *The inactivation of trypsin, III.* 25, 54
J. gen. physiol., **4**, 261.

J. H. Northrop (1922, 4), *Does the kinetics of trypsin digestion depend on the formation of a compound between enzyme and substrate?* . 6, 36, 38, 54, 94
J. gen. physiol., **4**, 487.

J. H. Northrop (1922, 5), *The mechanism of the influence of acids and alkalies on the digestion of proteins by pepsin or trypsin* 17, 22
J. gen. physiol., **5**, 263.

J. H. Northrop (1924, 1), *The kinetics of trypsin digestion, I.* 22
J. gen. physiol., **6**, 239.

J. H. Northrop (1924, 2), *A test for diffusible ions* 22
J. gen. physiol., **6**, 337.

Page of Text on which reference is made.

J. H. NORTHROP (1924, 3), *The kinetics of trypsin digestion. V. Schutz's rule* 91
J. gen. physiol., 6, 723.

J. H. NORTHROP (1930), *Crystalline pepsin* 170
J. gen. physiol., **13**, 739.

J. H. NORTHROP and H. S. SIMMS (1928), *The effect of the hydrogen ion concentration on the rate of hydrolysis of glycyl-glycine, glycyl-leucine, glycyl-alanine, glycyl-asparagine, glycyl-aspartic acid, and biuret base by erepsin* 23, 36
J. gen. physiol., **12**, 313.

K. NOSAKA (1928, 1), *Studien über die katalytische Spaltung des Wasserstoffsuperoxyds durch das Blut. I. Über die chemische Dynamik der Blutkatalase* 18, 79
J. Bioch., **8**, 275.

K. NOSAKA (1928, 2), *Studien über die katalytische Spaltung des Wasserstoffsuperoxyds durch das Blut. II. Über den Einfluss der Temperatur auf die Blutkatalase* 79
J. Bioch., **8**, 301.

K. NOSAKA (1928, 3), *Studien über die katalytische Spaltung des Wasserstoffsuperoxyds durch das Blut. IV. Über die sogenannte Hitzeaktivierung und den Einfluss einiger organischer Substanzen auf die Blutkörperchenkatalase* 71, 79
J. Bioch., **8**, 331.

E. OHLSSON (1921), *Die Abhangigkeit der Wirkung der Succinodehydrase von der Wasserstoffionen-Konzentration* 18, 168
Skand. Arch. Physiol., **41**, 77.

E. OHLSSON (1930), *Über die beiden Komponenten der Malzdiastase, etc.* . 116
Zeits. physiol. Chem., **198**, 17.

E. OHLSSON and T. SWAETICHIN (1929), *Etude sur la takadiastase* . . 26
Bull. Soc. Chim. Biol., 11.

K. OHTA (1914), *Darstellung von Eiweissfreiem Emulsin* 93
Bioch. Zts., **58**, 329.

P. OLITZKY and L. BOEZ (1927), *Studies on the physical and chemical properties of the virus of foot-and-mouth disease. II. Cataphoresis and filtration* 180
J. Exp. Med., **45**, 693.

U. OLSSON (1921), *Über Vergiftung der Amylase durch Schwermetalle und organische Stoffe* 152
Zts. physiol. Chem., **114**, 51.

U. OLSSON (1925), *Dissertation, Stockholm,* quoted by Myrbäck (1926, 2), p. 78 . . . . 151, 155

N. ONODERA (1915), *On the effects of various substances upon the urease of soy-bean* 146
Bioch. J., **9**, 544.

H. ONSLOW (1915), *A contribution to our knowledge of the chemistry of coat colour in animals and of dominant and recessive whiteness* . . . 163
Proc. Roy. Soc., **89**B, 36.

M. W. ONSLOW (1920), *Oxidizing enzymes. II. The nature of the enzymes associated with certain direct oxidizing systems in plants* . . . 191
Bioch. J., **14**, 535.

C. OPPENHEIMER (1910), *Die Fermente und Ihre Wirkungen.* (Fourth edition) 77

C. OPPENHEIMER (1925), *Die Fermente und Ihre Wirkungen.* (Fifth edition). 185

C. O'SULLIVAN and F. W. TOMPSON (1890), *Invertase ; a contribution to the history of an enzyme or unorganised ferment* 9
J. Chem. Soc., **57**, 834.

L. PASTEUR (1858), *Mémoire sur la fermentation de l'acide tartrique* . . 8
C. R. Ac. Sc., **46**, 615.

L. PASTEUR (1871), *Note sur un mémoire de M. Liebig, relatif aux fermentations* 8
C. R. Ac. Sc., **73**, 1419.

Page of Text on which reference is made.

A. PAYEN and J. PERSOZ (1833), *Mémoire sur la diastase, les principaux produits de ses réactions et leurs applications aux arts industriels* . . 8
Ann. Chim. Phys., **53**, 73.

C. PEKELHARING and W. RINGER (1911), *Zur elektrischen Überführung des Pepsins* 71
Zts. physiol. Chem., **75**, 282.

A. PICTET and R. SALZMANN (1924), *Sur la trihexosane* 119
Helv. Chim. Act., **7**, 934.

PIN YIN YI (1920), *Berichte d. deutsch. Pharm. Ges.*, 30, 178 (quoted by Euler, "*Die Enzyme, II.*," p. 335, 1927) 106

L. PINCUSSEN (1924, 1), *Fermente und Licht, V* 71
Bioch. Zts., **144**, 372.

L. PINCUSSEN (1924, 2, 3), *Fermente und Licht, VI., VII. Die Beeinflussung von Fermentwirkungen durch Jodsalze unter Bestrahlung, I., II.* 72
Bioch. Zts., **152**, 406, 416.

L. PINCUSSEN (1926), *Über den Einfluss der Reinigung von Fermenten auf ihre Empfindlichkeit gegenüber der Strahlung* 72
Fermentforschung, **8**, 181.

L. PLANCHE (1810), *Note sur la sophistication de la resine de jalap et les moyens de la connaitre* 8
Bull. Pharm. **82**, 578.

L. PLANCHE (1820), *Expériences sur les substances qui développent la couleur bleue dans la résine de gaïac* 8
Journ. de Pharm., **6**, 16.

B. S. PLATT and E. R. DAWSON (1925), *Factors influencing the action of pancreatic lipase* 15, 134
Bioch. J., **19**, 860.

A. E. PORTER (1916), *The distribution of esterases in the animal body*. . 102
Bioch. J., **10**, 523.

H. POTTEVIN (1903), *Influence de la configuration stéréochimique des glucosides sur l'activité des diastases hydrolytiques* 95
Ann. Inst. Past., **17**, 31.

H. PRINGSHEIM and A. BEISER (1924), *Über ein Komplement der Amylasen und das Grenzdextrin* 140
Bioch. Zts., **148**, 336.

H. PRINGSHEIM and A. GENIN (1924), *Über die fermentative Spaltung des Salepmannans* 16, 116
Zts. physiol. Chem., **140**, 299.

H. PRINGSHEIM and G. KOHN (1924), *Zur Kenntnis des Inulins und der Inulase* 17
Zts. physiol. Chem., **133**, 80.

H. PRINGSHEIM and J. LEIBOWITZ (1923), *Über Cellobiase und Lichenase* . 95
Zts. physiol. Chem., **131**, 262.

H. PRINGSHEIM and J. LEIBOWITZ (1925, 1), *Über die Maltase des Gerstenmalzes* 16, 17, 119
Bioch. Zts., **161**, 456.

H. PRINGSHEIM and J. LEIBOWITZ (1925, 2), *Über die Beziehung zwischen optischem Drehungsvermögen und Struktur in der Polysaccharid-Chemie* 19, 131
Ber., **58**, 2808.

H. PRINGSHEIM and G. OTTO (1926), *Über das Komplement der Amylasen* . 140
Bioch. Zts., **173**, 399.

H. PRINGSHEIM and E. SCHAPIRO (1926), *Über den fermentativen Abbau der Stärke durch "Biolase"* 17, 117
Ber., **59**, 996.

H. PRINGSHEIM and K. SEIFERT (1923), *Über die fermentative Spaltung des Lichenins* 17
Zts. physiol. Chem., **128**, 284.

H. PRINGSHEIM and M. WINTER (1926), *Über das Komplement der Amylasen und die Zucker-Eiweisskondensation* 140
Bioch. Zts., **177**, 406.

Page of Text on which reference is made.

T. PURDIE and J. C. IRVINE (1904), *The stereoisomeric tetramethyl methylglucosides and tetramethylglucose* 98
J. Chem. Soc., **85**, 1049.

J. H. QUASTEL (1926), *Dehydrogenations produced by resting bacteria. IV. A theory of the mechanism of oxidations and reductions in vivo* . 125, 183
Bioch. J., **20**, 166.

J. H. QUASTEL, M. STEPHENSON, and M. WHETHAM (1925), *Some reactions of resting bacteria in relation to anaerobic growth* 190
Bioch. J., **19**, 304.

J. H. QUASTEL and M. D. WHETHAM (1924), *The equilibrium existing between succinic, malic and fumaric acids in the presence of resting bacteria* 18, 66, 190
Bioch. J., **18**, 519.

J. H. QUASTEL and M. D. WHETHAM (1925, 1), *Dehydrogenations produced by resting bacteria, I.* 43, 60
Bioch. J., **19**, 520.

J. H. QUASTEL and M. D. WHETHAM (1925, 2), *Dehydrogenations produced by resting bacteria, II.* 60, 125
Bioch. J., **19**, 645.

J. H. QUASTEL and W. R. WOOLDRIDGE (1927, 1), *The effects of chemical and physical changes in the environment on resting bacteria* . 125, 152, 153, 190
Bioch. J., **21**, 148.

J. H. QUASTEL and W. R. WOOLDRIDGE (1927, 2), *Experiments on bacteria in relation to the mechanism of enzyme action* . . 62, 125 152, 153, 190
Bioch. J., **21**, 1224.

J. H. QUASTEL and W. R. WOOLDRIDGE (1928), *Some properties of dehydrogenating enzymes of bacteria* 61, 125, 190
Bioch. J., **22**, 689.

J. H. QUASTEL and B. WOOLF (1926), *The equilibrium between l-aspartic acid, fumaric acid and ammonia in presence of resting bacteria* . . 128
Bioch. J., **20**, 545.

H. S. RAPER (1928), *The aerobic oxidases* 191
Physiol. Reviews, **8**, 245.

J. REYNOLDS GREEN (1897), *On the action of light on diastase and its biological significance* 72
Proc. Roy. Soc., **61**, 25.

O. RIESSER (1906), *Zur Kentniss der optischen Isomeren des Arginins und Ornithins* 107
Zts. physiol. Chem., **49**, 210.

C. RIMINGTON (1927, 1), *Phosphorylation of proteins*. 105
Bioch. J., **21**, 272.

C. RIMINGTON (1927, 2), *The phosphorus of caseinogen. II. The constitution of phosphopeptone* 105, 106
Bioch. J., **21**, 1187.

R. ROBINSON (1927), *An aspect of the biochemistry of the sugars* . . 188
Nature, **120**, 44.

ROBIQUET (1838), Mentioned in: "*Extrait du procès verbal de la société de pharmacie de Paris, séance du 7 mars, 1838*" 8

ROBIQUET and BOUTRON-CHALARD (1830), *Nouvelles expériences sur les amandes amères* 8
Ann. Chim. Phys., **44**, 352.

E. W. ROCKWOOD (1924), *The mechanism of the action of amino promoters upon enzymes* 139, 147
J. Amer. Chem. Soc., **46**, 1641.

E. W. ROCKWOOD and W. J. HUSA (1923), *Studies on enzyme action. The relationship between chemical structure of certain compounds and their effect upon the activity of urease* 147
J. Amer. Chem. Soc., **45**, 2678.

ROHDEWALD (1927). Quoted by Kuhn and Munch (1927) 100

Page of Text on which reference is made.

E. RONA (1920), *Über die Wirksamkeit der Fermente unter abnormen Bedingungen und über die angebliche Aldehydnatur der Enzyme* . . 155
Bioch. Zts., **109**, 279.

P. RONA and R. AMMON (1927) *Zur stereochemischen Spezifität der Lipasen und Beiträge zur Giftwirkung an den fettspaltenden Fermenten* 74, 88, 103, 104
Bioch. Zts., **181**, 49.

P. RONA and E. BACH (1920), *Beiträge zum Studium der Giftwirkung. Über die Wirkung des Atoxyls auf Serumlipase* 160
Bioch. Zts., **111**, 166.

P. RONA and E. BACH (1921), *Beiträge zum Studium der Giftwirkung. Über die Wirkung des m- und p-Nitrophenols auf Invertase* . . . 152
Bioch. Zts., **118**, 232.

P. RONA and E. BLOCH (1921), *Beiträge zum Studium der Giftwirkung. Über die Wirkung des Chinins auf Invertase* 155
Bioch. Zts., **118**, 185.

P. RONA, C. VAN EWEYK, and M. TENNENBAUM (1924), *Über die Wirkung der Alkaloide aus der Atropin- Cocain- und Morphingruppe auf die Hefe-Invertase* 155
Bioch. Zts., **144**, 490.

P. RONA and P. GYORGY (1920), *Zur Kenntnis der Urease. Zugleich ein Beitrag zum Studium der Giftwirkungen* 162
Bioch. Zts., **111**, 115.

P. RONA and K. GYOTOKU (1926), *Weitere Beiträge zur Lipasevergiftung durch Chinin und Atoxyl* 160
Bioch. Zts., **167**, 171.

P. RONA and HAAS (1923), *Über die Wirkung des Chinins und Atoxyls auf die Nierenlipase* 161
Bioch. Zts., **141**, 222.

P. RONA and HEFTER (1930), *Über die gleichzeitige Wirkung von Speichel-, Pankreas-, und Malzamylase auf Starke* 119
Bioch. Zts., **217**, 113.

P. RONA and R. ITELSOHN-SCHECHTER (1928), *Über die Spaltung des Milchsäureethylesters durch Leberesterase* 103
Bioch. Zts., **203**, 293.

P. RONA and H. KLEINMANN (1926), *Untersuchungen über die Beziehung zwischen Substratdispersität und Fermentwirkung Mittels einer neuen Methode zur Bestimmung der fermentativen Fettspaltung* . . . 137
Bioch. Zts. **174**, 18.

P. RONA and A. LASNITZKY (1924), *Eine Methode zur Bestimmung der Lipase in Körperflüssigkeiten und im Gewebe* 74
Bioch. Zts., **152**, 504.

P. RONA and L. MICHAELIS (1913), *Die Wirkungsbedingungen der Maltase aus Bierhefe, II.* 15
Bioch. Zts., **58**, 148.

P. RONA and R. PAVLOVIČ (1922), *Beitrage zum Studium der Giftwirkung. Über die Wirkung des Chinins und Atoxyls auf Leberlipase* . . . 160
Bioch. Zts., **130**, 225.

P. RONA and R. PAVLOVIČ (1923), *Über die Wirkung des Chinins und Atoxyls auf Pancreaslipase* 160
Bioch. Zts., **134**, 108.

P. RONA and H. PETOW (1924), *Weitere Untersuchungen über die Giftempfindlichkeit von Lipasen verschiedener Herkünft* 160
Bioch. Zts., **146**, 144.

P. RONA and D. REINECKE (1921), *Beiträge zum Studium der Giftwirkung. Über die Wirkung des Chinins auf Serumlipase* 160
Bioch. Zts., **118**, 213.

P. RONA and M. TAKATA (1923), *Über Gelbildung in Chinin- und Eucupinlosungen* 161
Bioch. Zts., **134**, 97.

L. ROSENTHALER (1908), *Durch Enzyme bewirkte asymmetrische Synthesen, I.* 128
Bioch. Zts., **14**, 238.

Page of Text on which reference is made.

L. ROSENTHALER (1909), *Durch Enzyme bewirkte asymmetrische Synthesen, II.* Bioch. Zts., **17,** 257. 128

L. ROSENTHALER (1913), *Zur Kenntnis emulsinartiger Enzyme* . . . Bioch. Zts., **50,** 486. 128

L. ROSENTHALER (1922), *σ-Emulsin (Oxynitrilese), δ-Emulsin (Oxynitrilase) Carboligasen* . . . . Bioch. Zts., **128,** 606. 128

P. ROTHMANN, D. WIDENER, and W. DAVISON (1927). Quoted by Waksman and Davison, *Enzymes* 16

SAMEC (1927), *Kolloidchemie der Stärke*, p. 283 116

SAMEC and KNEZ (1925), *Sur l'hydrolyse enzymatique des amylophosphates naturels et synthétiques* . . . . C. R. Ac. Sc., **181,** 532. 105

M. SANDBERG and E. BRAND (1925), *On papain lipase* . . . . . J. Biol. Chem., **64,** 59. 15

N. P. SCHEPOWALNIKOW (1899), *Die Physiologie des Darmsaftes* (Russian) . Maly's Jahresber, **29,** 378. 141

N. SCHLESINGER (1926), *Gleichgewichtsverschiebungen durch stoffe, die gleichzeitig katalytisch wirken, I.* . . . . Ber., **59,** 1965 3

N. SCHLESINGER (1927), *Gleichgewichtsverschiebungen, etc., II. Experimenteller Teil* . . . . Ber., **60,** 1479. 3

H. SCHLUBACH and G. RAUCHALLES (1925), *Die Spaltung des γ-Methylfructosides durch Saccharasen. Zur Konfiguration des Rohrzuckers* . Ber., **58,** 1842. 99, 131

SCHMIDT (1901), *Dissertation.* Quoted in Oppenheimer: "*Die Fermente*" (1925), p. 235 102

G. SCHMIDT (1929), *Über fermentative Desaminierung im Muskel* . . Zts. physiol. Chem., **179,** 243. 17

K. SCHNEIDER. *Saccharase.* In Oppenheimer and Pincussen: "*Die Methodik der Fermente,*" p. 765 . . . . 173, 178

E. SCHUNCK (1854), *Über die Einwirkung des Krappferments auf Zucker* . J. Prakt. Chem., **63,** 222. 188

A. SCHÜRMEYER (1925, 1), *Über Ionenantagonismus bei den Systemen Invertase-Eiweiss und Invertase-Lecithin, I.* . . . . . . Pflüg. Arch., **208,** 595. 158, 159

A. SCHÜRMEYER (1925, 2), *Über Ionenantagonismus bei den Systemen Invertase-Eiweiss und Invertase-Lecithin, II.* . . . . . . Pflüg. Arch., **210,** 755. 158, 159

E. SCHUTZ (1885), *Eine Methode zur Bestimmung der relativen Pepsinmenge* Zts. physiol. Chem., **9,** 577. 91

J. SCHUTZ (1900), *Zur Kenntnis der quantitativen Pepsinwirkung* . . Zts. physiol. Chem., **30,** 1. 91

T. SCHWANN (1836), *Über das Wesen der Verdauungsprocessus* . . . Müll. Arch., 1836, 90. 8

F. SELIGSOHN (1926), *Katalase, I.* . . . . Bioch. Zts., **168,** 457. 71

T. N. SETH (1924), *The activation of pancreatic juice by enterokinase* . . Bioch. J., **18,** 1401. 141

H. C. SHERMAN, M. L. CALDWELL, and M. ADAMS (1926), *Enzyme purification by adsorption: an investigation of pancreatic amylase* . . J. Amer. Chem. Soc., **48,** 2947. 167

H. C. SHERMAN, M. L. CALDWELL, and M. ADAMS (1928, 1), *The interrelation of hydrogen ion activity and concentration of salt in the activation of pancreatic amylase* . . . . J. Amer. Chem. Soc., **50,** 2529. 138, 139

Page of Text on which reference is made.

H. C. SHERMAN, M. L. CALDWELL, and M. ADAMS (1928, 2), *The influence of concentration of neutral salts on the activation of pancreatic amylase* . 138, 139
J. Amer. Chem. Soc., **50,** 2535.

H. C. SHERMAN, M. L. CALDWELL, and M. ADAMS (1928, 3), *A quantitative comparison of the influence of neutral salts on the activity of pancreatic amylase* 138, 139
J. Amer. Chem. Soc., **50,** 2538.

H. C. SHERMAN and A. P. TANBERG (1916), *Experiments upon the amylase of Aspergillus oryzae* 166
J. Amer. Chem. Soc., **38,** 1638.

H. C. SHERMAN, A. W. THOMAS, and M. E. BALDWIN (1919), *Hydrogen ion concentration and amylase* 17
J. Amer. Chem. Soc., **41,** 231.

E. SIEBURG and G. MORDHORST (1919), *Über die Verbreitung von Fermenten im tierischen Organismus, die Gerbsaüre und verwandte Stoffe Spalten* . 102
Bioch. Zts., **100,** 204.

K. SJÖBERG (1921), *Enzymatische Untersuchungen an einigen Grünalgen* . 17
Fermentforschung, **4,** 97.

K. SJÖBERG (1922), *Über die Bildung und das Verhalten der Amylase in lebenden Pflanzen* 17
Bioch. Zts., **133,** 218.

K. SJÖBERG (1924), *Über einige neue Produkte der enzymatischen Spaltung der Stärke* 140
Ber., **57,** 1251.

K. SJÖBERG and E. ERIKSSON (1924), *Amylase* 34, 36
Zts. physiol. Chem., **139,** 118.

I. A. SMORODINZEW (1923), *Über die Wirkung des Histozyms auf die Homologen der Hippursaüre* 106, 107
Zts. physiol. Chem., **124,** 123.

S. P. L. SÖRENSEN (1909), *Enzymstudien. II. Über die Messung und die Bedeutung der Wasserstoffionenkonzentration bei enzymatischen Prozessen* 9, 18, 29
Bioch. Zts., **21,** 131.

L. SPALLANZANI (1783), *Dissertazioni di fisica animale e vegetabile* (Modena) 67

R. STANKOVIC, V. ARNOVLJEVIC, and P. MATAVULJ (1929), *Enzymatische Hydrolyse des Keratins mit dem Kropfsaft des Astur palumbarius und Vultur monachus* 18, 189
Zts. physiol. Chem., **181,** 291.

M. STEPHENSON (1912), *On the nature of animal lactase* 101
Bioch. J., 250.

M. STEPHENSON (1928), *On lactic dehydrogenase. A cell-free enzyme preparation obtained from bacteria* 61, 126, 156, 191
Bioch. J., **22,** 605.

M. STEPHENSON (1930), "*Bacterial Metabolism.*" (This series) . . 62

L. STERN (1927), *Die Beziehung des katalasesystems zu den Oxydationsvorgängen in den Tiergeweben* 163, 191
Bioch. Zts., **182,** 139.

J. B. SUMNER (1926, 1), *The isolation and crystallisation of the enzyme urease.* 169
J. Biol. Chem., **69,** 435.

J. B. SUMNER (1926, 2), *The recrystallisation of urease* 169
J. Biol. Chem., **70,** 97.

J. B. SUMNER and D. B. HAND (1928), *Crystalline urease* . . 148, 166, 170, 176
J. Biol. Chem., **76,** 149.

A. SZENT-GYÖRGYI (1924), *Über den Mechanismus der Succin- und Paraphenylendiamin-Oxydation. Ein Beitrag zur Theorie der Zellatmung* 156
Bioch. Zts., **150,** 195.

Page of Text on which reference is made.

A. SZENT-GYÖRGYI (1928), *Observations on the function of peroxidase systems and the chemistry of the adrenal cortex. Description of a new carbohydrate derivative* 191
Bioch. J., **22,** 1387.

TAMMANN (1895), *Zur Wirkung ungeformter Fermente* 69
Zts. physikal. Chem., **18,** 426.

H. v. TAPPEINER (1908), *Über die sensibilisierende Wirkung fluoreszierender Stoffe auf Hefe und Hefepressaft* 73
Bioch. Zts., **8,** 47.

H. v. TAPPEINER and A. JODLBAUER (1925), *Die Beteiligung des Sauerstoffs bei der Wirkung fluoreszierendes Stoffe* 72
D. Arch. Klin. Med., **82,** 520.

TAUBMANN (1925), *Auxowirkung eiweissfreier Kolloide auf die Harnstoffspaltung durch Soja-Urease* 148
Bioch. Zts., **157,** 98.

J. TEMMINCK GROLL (1924), *Influence de la concentrations d'ions d'hydrogéne sur l'action de quelques amylases* 16, 17
Arch. Néerl. Physiol., **9,** 520.

TER MEULEN (1905), *Recherches experimentales sur la nature des sucres de quelques glucosides* 64, 79
Rec. Trav. Chim. Pays. Bas, **24,** 444.

TERROINE (1910, 4), *Fettspaltung durch Pankreassaft, I., II.* . . . 53, 65, 136
Bioch. Zts., **23,** 404, 429.

THOMAS (1913), *Über die Herkunft des Kreatins im tierischen Organismus* . 107
Zts. physiol. Chem., **88,** 465.

T. THUNBERG (1905), *Der Gasaustausch einiger niederer Tiere in seiner Abhängigkeit vom Sauerstoffpartiardruck* 37
Skand. Arch. Physiol., **17,** 133.

T. THUNBERG (1917), *Zur Kenntnis der Einwirkung Tierischer Gewebe auf Methylenblau* 9, 126
Skand. Arch. Physiol., **35,** 163.

T. THUNBERG (1929), *Über das Vorkommen einer Citrico-Dehydrogenase in Gurkensamen und ihre Verwertung fur eine hochempfindliche biologische Farbenreaktion auf Citronensaure* 37
Bioch. Zts., **206,** 109.

S. THURLOW (1925), *Studies on xanthine oxidase. IV. Relation of xanthine oxidase and similar oxidising systems to Bach's oxygenase* . . . 189
Bioch. J., **19,** 175.

S. TODA (1926), *Über Wasserstoffaktivierung durch Eisen* 157
Bioch. Zts., **172,** 34.

H. UCKO and H. W. BANSI (1926), *Über Peroxydase. II. Das Verhalten der peroxydatischen Fermente bei Variierung des Oxydationsobjekts* . 18
Zts. physiol. Chem., **157,** 214.

K. UEHARA (1928), *Pepsin, I.* 71
Bioch. Zts., **195,** 87.

S. UTZINO (1928, 1), *Über die Wirkung der proteolytischen Fermente auf die Benzoyl- und Phthalyl-Derivate der Polypeptide. I. Über die Wirkung des Darmerepsins und der Hefeprotease auf Phthalylglycin und Phthalyldiglycylglycin* 113
J. Bioch., **9,** 453.

S. UTZINO (1928, 2), *Über die Wirkung etc. II. Über die Wirkung der Gewebsproteasen auf Benzoyl- und Phthalylglycylglycin* 113
J. Bioch., **9,** 465.

S. UTZINO (1928, 3), *Über die Wirkung etc. III. Über die Wirkung der Pankreasproteasen auf Benzoyl- und Phthalylglycin* 113
J. Bioch., **9,** 483.

D. D. van SLYKE and G. E. CULLEN (1914). *The mode of action of urease and of enzymes in general* 36, 41, 66
J. Biol. Chem., **19,** 141.

Page of Text on which reference is made.

L. VELLUZ (1927), *Recherches sur l'action inhibitrice des acides gras et, en particulier, les acides non saturés, sur les phenoménes diastasiques* . 159
Bull. Soc. Chim. Biol., **9**, 483.

H. M. VERNON (1914), *The activation of trypsinogen* 141
Bioch. J., **8**, 494.

A. J. VIRTANEN and H. KARSTRÖM (1928), *Über die Milchsäuregärung, V.* . 18
Zts. physiol. Chem., **174**, 1.

*E. WALDSCHMIDT-LEITZ (1924), *Über Enterokinase und die tryptische Wirkung der Pankreasdrüse* 141
Zts. physiol. Chem., **132**, 181.

*E. WALDSCHMIDT-LEITZ (1925), *Zur Kenntnis der Enterokinase* . . 141
Zts. physiol. Chem., **142**, 217.

E. WALDSCHMIDT-LEITZ (1926), "*Die Enzyme*" (Braunschweig), p. 80 108, 168, 172

E. WALDSCHMIDT-LEITZ, A. K. BALLS, and J. WALDSCHMIDT-GRASER (1929), *Über Dipeptidase und Polypeptidase aus Darm-Schleimhaut*. . . 17, 108
Ber., **62**, 956.

E. WALDSCHMIDT-LEITZ, I. BEK, and J. KAHN (1929), *Über die Aktivierung der Proteolyse in tierischen Organen und ihre Bedeutung für den Stoffwechsel bösartiger Geschwülste* 108
Naturwiss., **17**, 85.

*E. WALDSCHMIDT-LEITZ and W. DEUTSCH (1927), *Über die Proteolytischen Enzyme der Milz* 188
Zts. physiol. Chem., **167**, 285.

*E. WALDSCHMIDT-LEITZ and A. HARTENECK (1925), *Zur Kenntnis der Spontanen Aktivierung des Trypsins* 143
Zts. physiol. Chem., **149**, 221.

*E. WALDSCHMIDT-LEITZ and W. KLEIN (1928), *Über Spezifität und Wirkungsweise von Erepsin, Trypsin und Trypsin-Kinase*. . . 113
Ber., **61**, 640.

*E. WALDSCHMIDT-LEITZ, W. KLEIN, and A. SCHAFFNER (1928), *Über die strukturellen Voraussetzungen der spezifischen Spaltbarkeit proteolytischer Substrate. Zur Spezifität von Trypsin, Trypsin-Kinase und Darm-Erepsin* 112, 113
Ber., **61**, 2092.

*E. WALDSCHMIDT-LEITZ and A. PURR (1929), *Spezifität tierische Proteasen. XVII. Proteinase und Carboxypolypeptidase aus Pankreas* . . . 108, 109
Ber., **62**, 2217

E. WALDSCHMIDT-LEITZ, A. PURR, AND A. K. BALLS (1930), *Über der natürlichen Aktivator der Katheptischen Enzyme* 144
Naturwissenschaften, **28**, 644.

E. WALDSCHMIDT-LEITZ, A. SCHÄFFNER, and W. GRASSMANN (1926), *Über die Struktur des Clupeins* 29
Zts. physiol. Chem., **156**, 68.

*E. WALDSCHMIDT-LEITZ, A. SCHÄFFNER, H. SCHLATTER, and W. KLEIN (1928), *Zur Spezifität von Pankreas-Trypsin und Darm-Erepsin* . . 113
Ber., **61**, 299.

*E. WALDSCHMIDT-LEITZ and H. SCHLATTER (1928), *Zur Stereochemischen Spezifität proteolytischer Enzyme* 111
Naturwissenschaften, **16**, 1026.

O. WARBURG (1927), *Über die Wirkung von Kohlenoxyd und Stickoxyd auf Atmung und Gärung*. 37, 157
Bioch. Zts., **189**, 355.

O. WARBURG and KUBOWITZ (1929), *Atmung bei sehr kleinen Sauerstoffdrucken* 37, 127
Bioch. Zts., **214**, 5.

H. H. WEBER and R. AMMON (1929), *Die stereochemische Spezifität der Leber- und Pankreaslipase* 35, 85, 86
Bioch. Zts., **204**, 197.

R. WEIDENHAGEN (1928), *Zur Frage der enzymatischen Rohrzuckerspaltung* 100
Naturwissenschaften, **16**, 654.

Page of Text on which reference is made.

E. M. P. WIDMARK (1922), *Studien über die Succinodehydrogenase* . . 37
Skand. Arch. Physiol., **41**, 200.

E. M. P. WIDMARK and C. A. JEPPSON (1922), *Ein definierter organischer Katalysator mit Wasserstoffionenoptimum* 4
Skand. Arch. Physiol., **42**, 43.

H. WIELAND and G. F. FISCHER (1926), *Zur Frage der katalytischen Dehydrierung* 192
Ber., **59**, 1180.

H. WIELAND and B. ROSENFELD (1929), *Über den Mechanismus der Oxydationsvorgänge. XXI. Über dehydrierende enzyme der Milch* . . 123
Ann. Chem., **477**, 32.

H. WIELAND and H. SUTTER (1928), *Einiges über Oxidasen und Peroxydasen* 155, 166, 192
Ber., **61**, 1060.

H. WIELAND and H. SUTTER (1930), *Beiträge zur Wirkungsweise von Oxydasen und Peroxydasen. XXII. Mitteilung, Über den Mechanismus der Oxydationsvorgänge* 18
Ber., **63**, 67.

V. B. WIGGLESWORTH (1927), *Digestion in the cockroach. II. The digestion of carbohydrates* 13, 15, 16
Bioch. J., **21**, 797.

*R. WILLSTÄTTER (1922), *Über Isolierung von Enzymen* . . . 29, 171, 176
Ber., **55**, 3601.

*R. WILLSTÄTTER (1926), *Über Sauerstoff-Ubertrügung in der lebenden Zelle*. 190
Ber., **59**, 1871.

*R. WILLSTÄTTER and E. BAMANN (1928), *Über Magenlipase* 134, 166, 172, 174, 175
Zts. physiol. Chem., **173**, 17.

*R. WILLSTÄTTER, E. BAMANN, and J. WALDSCHMIDT-GRASER (1928), *Über die Konfigurationsspezifität der Esterasen in verschiedenen Reinheitsgraden* 104
Zts. physiol. Chem., **173**, 155.

*R. WILLSTÄTTER and W. CSANYI (1921) *Zur Kenntnis des Emulsins*. . 16, 169
Zts. physiol. Chem., **117**, 172.

*R. WILLSTÄTTER, J. GRASER, and R. KUHN (1922), *Zur Kenntnis des Invertins* 21, 175
Zts. physiol. Chem., **123**, 1.

*R. WILLSTÄTTER and W. GRASSMANN (1924), *Über die Aktivierung des Papains durch Blausäure* 108, 115, 144
Zts. physiol. Chem., **138**, 184.

*R. WILLSTÄTTER and W. GRASSMANN (1926), *Über die Proteasen der Hefe* . 17
Zts. physiol. Chem., **153**, 250.

*R. WILLSTÄTTER, W. GRASSMANN, and O. AMBROS (1926), *Blausäure Aktivierung und Hemmung pflanzlicher Proteasen* . . . 18, 23, 144
Zts. physiol. Chem., **151**, 286.

*R. WILLSTÄTTER, F. HAUROWITZ, and F. MEMMEN (1924), *Zur Spezifität der Lipasen aus verschiedenen Organen* 13
Zts. physiol. Chem., **140**, 203.

*R. WILLSTÄTTER and H. KRAUT (1923, 1), *Zur Kenntnis der Tonerdehydrate* 171
Ber., **56**, 149.

*R. WILLSTÄTTER and H. KRAUT (1923, 2), *Über ein Tonerdegel der Formel* $Al(OH)_3$ 172
Ber., **56**, 1117.

*R. WILLSTÄTTER and H. KRAUT (1924, 1), *Über Wasserarme Tonerdehydrate* 172
Ber., **57**, 58.

*R. WILLSTÄTTER and H. KRAUT (1924, 2), *Über die Hydroxyde und ihre Hydrate in den verschiedenen Tonerdegelen* 172
Ber., **57**, 1082.

*R. WILLSTÄTTER and R. KUHN (1921, 1), *Bemerkungen über die Elution von Saccharase und Maltase aus ihren Adsorbaten* 29
Zts. physiol. Chem., **116**, 53.

Page of Text on which reference is made.

*R. Willstätter and R. Kuhn (1923), *Über Masseinheiten der Enzyme* . 167
Ber., **56**, 509.

*R. Willstätter, R. Kuhn, and E. Bamann (1928), *Über asymmetrisch Esterhydrolyse durch Enzyme. (I. Mitteilung)* 35, 85-88, 104
Ber., **61**, 886.

*R. Willstätter, R. Kuhn, O. Lind, F. Memmen (1927), *Über Hemmung der Leberesterase durch ketocarbonsaüreester* 53
Zts. physiol. Chem., **167**, 303.

*R. Willstätter, R. Kuhn, and H. Sobotka (1923), *Über die einheitliche Natur der β-Glukosidase und des Emulsins* 36, 44
Zts. physiol. Chem., **129**, 33.

*R. Willstätter, R. Kuhn and H. Sobotka (1924), *Über die relative Spezifität der Hefemaltase* 35, 36, 45
Zts. physiol. Chem., **134**, 224.

*R. Willstätter and H. Kumagawa (1925), *Über Taka-Esterase; Vergleich mit Pankreaslipase und Leberesterase* 15
Zts. physiol. Chem., **146**, 151.

*R. Willstätter and C. D. Lowry (1925), *Invertinverminderung in der Hefe* 122
Zts. physiol. Chem., **150**, 287.

*R. Willstätter, C. D. Lowry, and K. Schneider (1925), *Invertinanreicherung in der Hefe* 172
Zts. physiol. Chem., **146**, 158.

*R. Willstätter and F. Memmen (1923), *Zur stalagmometrischen Bestimmung der Lipatischen Tributyrinhydrolyse* 134, 135, 168
Zts. physiol. Chem., **129**, 1.

*R. Willstätter and F. Memmen (1924, 1), *Über die Wirkung der Pankreaslipase auf verschiedene Substrate* 35, 135
Zts. physiol. Chem., **133**, 229.

*R. Willstätter and F. Memmen (1924, 2), *Vergleich von Leberesterase mit Pankreaslipase; über die stereochemische Spezifität der lipasen* . 103, 161
Zts. physiol. Chem., **138**, 216.

*R. Willstätter and G. Oppenheimer (1922), *Über Laktasegehalt und Gärvermögen von Milchzuckerhefen* 121
Zts. physiol. Chem., **134**, 224.

*R. Willstätter and A. Pollinger (1923), *Über Peroxydase*
Ann. Chem., **430**, 290. 1, 166, 167, 169-172, 179

*R. Willstätter and F. Racke (1921), *Zur Kenntnis des Invertins* . . 160
Ann. Chem., **425**, 1.

*R. Willstätter and K. Schneider (1925), *Zur Kenntnis des Invertins* . 166
Zts. physiol. Chem., **142**, 257.

*R. Willstätter, K. Schneider and E. Wenzel (1926), *Zur Kenntnis des Invertins* 172, 175
Zts. physiol. Chem., **151**, 1.

*R. Willstätter and Sobotka (1922), *Vergleich von α- und β-glucose in der Gärung* 122
Zts. physiol. Chem., **123**, 170.

*R. Willstätter and W. Steibelt (1920), *Bestimmung der Maltase in der Hefe* 16
Zts. physiol. Chem., **111**, 157.

*R. Willstätter and W. Steibelt (1921), *Über die Gärwirkung Maltasearmer Hefen* 121
Zts. physiol. Chem., **115**, 211.

R. Willstätter and A. Stoll (1913), *Untersuchungen über Chlorophyll.* (Springer.) 102

*R. Willstätter and A. Stoll (1918), *Über Peroxydase* 169, 172
Ann. Chem., **416**, 21.

*R. Willstätter and E. Waldschmidt-Leitz (1922), *Über Pankreaslipase* 29, 166, 168
Zts. physiol. Chem., **125**, 132.

Page of Text on which reference is made.

*R. Willstätter and E. Waldschmidt-Leitz (1924), *Über Ricinus Lipase* 26, 166, 168
Zts. physiol. Chem., **134**, 161.

*R. Willstätter, E. Waldschmidt-Leitz, and A. R. F. Hesse (1925), *Über das Adsorptionsverhalten der Pankreasamylase* 166
Zts. physiol. Chem., **142**, 14.

*R. Willstätter, E. Waldschmidt-Leitz, and F. Memmen (1923), *Bestimmung der pankreatischen Fettspaltung* 134, 135, 167
Zts. physiol. Chem., **125**, 93.

*R. Willstätter and H. Weber (1926, 1), *Zur quantitativen Bestimmung der Peroxydase* 37, 59, 66, 175, 177
Ann. Chem., **449**, 156.

*R. Willstätter and H. Weber (1926, 2), *Über Hemmung der Peroxydase durch Hydroperoxyd* 59
Ann. Chem., **449**, 175.

A. Wohl and E. Glimm (1910), *Zur Kenntnis der Amylase* 63, 64
Bioch. Zts., **27**, 349.

B. Woolf (1929), *Some enzymes of B. coli communis which act on fumaric acid* 37, 83, 126, 128, 184
Bioch. J., **23**, 472.

F. Wrede and O. Hettche (1927), *Über Thiocellobiose und Thiocellobioside* 96
Zts. physiol. Chem., **172**, 169.

Yamazaki (1921). (Cited by Nosaka, 1927) 79
Sci. Rep. Tohoku. Imp. Univ., 9.

M. Zeller and A. Jodlbauer (1908), *Die Sensibilisierung der Katalase* . 72, 73
Bioch. Zts., **8**, 84.

G. Zemplen (1926, 1), *Abbau der reduzierenden Biosen. III. Direkte Konstitutions-Ermittlung des Milchzuckers* 95
Ber., **59**, 2402.

G. Zemplen (1926, 2), *Abbau der reduzierenden Biosen. IV. Nachtrag zur Konstitution der Turanose und Melezitose* 131
Ber., **59**, 2539.

G. Zemplen (1927), *Abbau der reduzierenden Biosen. V. Konstitutions-Ermittlung der Melibiose und der Raffinosr* 131
Ber., **60**, 923.

S. S. Zilva (1914), *The rate of inactivation by heat of peroxidase in milk, I.* . 69
Bioch. J., **8**, 656.

SUBJECT INDEX.

14